Cold-Formed Steel Structures
to the AISI Specification

Civil and Environmental Engineering
A Series of Reference Books and Textbooks

Editor
Michael D. Meyer

Department of Civil and Environmental Engineering
Georgia Institute of Technology
Atlanta, Georgia

Additional Volumes in Production

Handbook of Pollution Control and Waste Minimization
Abbas Ghassemi

Introduction to Approximate Solution Techniques, Numerical Modeling, and Finite Element Methods
Victor N. Kaliakin

Cold-Formed Steel Structures to the AISI Specification

Gregory J. Hancock
University of Sydney
Sydney, New South Wales, Australia

Thomas M. Murray
Virginia Polytechnic Institute and State University
Blacksburg, Virginia

Duane S. Ellifritt
University of Florida
Gainesville, Florida

CRC Press
Taylor & Francis Group
Boca Raton London New York

CRC Press is an imprint of the
Taylor & Francis Group, an **informa** business

CRC Press
Taylor & Francis Group
6000 Broken Sound Parkway NW, Suite 300
Boca Raton, FL 33487-2742

First issued in paperback 2019

© 2001 by Taylor & Francis Group, LLC
CRC Press is an imprint of Taylor & Francis Group, an Informa business

No claim to original U.S. Government works

ISBN-13: 978-0-367-39706-7

Visit the Taylor & Francis Web site at
http://www.taylorandfrancis.com

and the CRC Press Web site at
http://www.crcpress.com

Preface

This book describes the structural behavior and design of cold-formed steel structural members, connections, and systems. Cold-formed members are often more difficult to design than conventional hot-rolled members because of their thinness and the unusual shapes that can be roll-formed. Design engineers familiar with hot-rolled design can sometimes find cold-formed design a daunting task. The main objective of this book is to provide an easy introduction to the behavior of cold-formed members and connections with simple worked examples. Engineers familiar with hot-rolled design should be able to make the step to cold-formed design more easily using this text. The book is in line with the latest edition of the American Iron and Steel Institute (AISI) Specification for the Design of Cold-Formed Steel Structural Members (1996), including the latest amendment (Supplement 1 1999) published in July 2000.

Preface

The book describes the main areas of cold-formed members, connections, and systems in routine use today. It presents the history and applications of cold-formed steel design in Chapter 1, cold-formed steel materials in Chapter 2, and buckling modes of thin-walled members in Chapter 3. These three chapters provide a basic understanding of how cold-formed steel design differs from hot-rolled steel design. Chapters 4 to 9 follow the approach of a more conventional text on structural steel design, with Chapter 4 on elements, Chapter 5 on flexural members, Chapter 6 on webs, Chapter 7 on compression members, Chapter 8 on combined compression and bending, and Chapter 9 on connections. Chapters 10 and 11 apply the member design methods to common systems in which cold-formed steel is used, roof and wall systems in Chapter 10 and steel storage racks in Chapter 11. Chapter 12 gives a glimpse into the future of cold-formed steel design where the direct strength method is introduced. This method based on the finite strip method of structural analysis introduced in Chapter 3 and currently under consideration by the AISI Specification Committee, gives a simpler and more direct approach to design than the current effective width approach. Residential framing is not covered in a separate chapter since the design methods in Chapters 4 to 9 are appropriate for members and connections used in wall framing and roof trusses.

The book has been based on a similar one, *Design of Cold-Formed Steel Structures*, written by the first author and published by the Australian Institute of Steel Construction in 1988, 1994, and 1998 with design to the Australian/New Zealand Standard AS/NZS 4600, which is similar to the AISI Specification. The Australian book is completely in SI units to the Australian standard, whereas this book is written in U.S. customary units to the AISI specification.

The worked examples in Chapters 4–8, 10 and 11 have all been programmed as MATHCAD^{TM} spreadsheets to ensure their accuracy. Hence, the numerical values quoted in these chapters are taken directly from MATHCAD^{TM} and have been rounded to the appropriate number of significant figures at each output. Rounding is therefore not included in the computation as it would be if the values were rounded progressively as is done in manual calculations. The final values quoted in the book at the end of the examples can therefore be compared directly with those obtained using other computer software.

The book has involved substantial work to convert it from the Australian version, with new Chapters 1 and 10 being written by the third and second authors, respectively. The text was typed by Gwenda McJannet, and the drawings were prepared by Mr. Ron Brew and Ms. Kim Pham, all at the University of Sydney. We are grateful to these people for their careful work. The first author is grateful for the advice on cold-formed steel design from Professor Teoman Peköz at Cornell University and Emeritus Professor Wei-Wen Yu at the University of Missouri–Rolla over many years.

The assistance of Dr. Ben Schafer at Johns Hopkins University with Chapter 12 is appreciated.

Gregory J. Hancock
Thomas M. Murray
Duane S. Ellifritt

Contents

Contents

Contents

Contents

Contents

Contents

Contents

1

Introduction

1.1 DEFINITION

Cold-formed steel products are just what the name connotes: products that are made by bending a flat sheet of steel at room temperature into a shape that will support more load than the flat sheet itself.

1.2 BRIEF HISTORY OF COLD-FORMED STEEL USAGE

While cold-formed steel products are used in automobile bodies, kitchen appliances, furniture, and hundreds of other domestic applications, the emphasis in this book is on *structural* members used for buildings.

Cold-formed structures have been produced and widely used in the United States for at least a century. Corrugated sheets for farm buildings, corrugated culverts, round grain bins, retaining walls, rails, and other structures have been around for most of the 20th century. Cold-formed steel for industrial and commercial buildings began about mid-20th century, and widespread usage of steel in residential buildings started in the latter two decades of the century.

1.3 THE DEVELOPMENT OF A DESIGN STANDARD

It is with structural steel for *buildings* that this book is concerned and this requires some widely accepted standard for design. But the design standard for hot-rolled steel, the American Institute of Steel Construction's Specification (Ref. 1.1), is not appropriate for cold-formed steel for several reasons. First, cold-formed sections, being thinner than hot-rolled sections, have different behavior and different modes of failure. Thin-walled sections are characterized by local instabilities that do *not* normally lead to failure, but are helped by postbuckling strength; hot-rolled sections rarely exhibit local buckling. The properties of cold-formed steel are altered by the forming process and residual stresses are significantly different from hot-rolled. Any design standard, then, must be particularly sensitive to these characteristics which are peculiar to cold-formed steel.

Fastening methods are different, too. Whereas hot-rolled steel members are usually connected with bolts or welds, light gauge sections may be connected with bolts, screws, puddle welds, pop rivets, mechanical seaming, and sometimes "clinching."

Second, the industry of cold-formed steel differs from that of hot-rolled steel in an important way: there is much less standardization of shapes in cold-formed steel. Rolling heavy structural sections involves a major investment in equipment. The handling of heavy billets, the need to reheat them to 2300°F, the heavy rolling stands capable of exerting great pressure on the billet, and the loading, stacking, and storage of the finished product all make the production of hot-rolled steel shapes a significant financial investment by the manufacturer. Thus, when a designer specifies a W21×62, for example, he can be assured that it will have the same dimensions, no matter what company makes it.

Conversely, all it takes to make a cold-formed structural shape is to take a flat sheet at room temperature and bend it. This may be as simple as a single person lifting a sheet onto a press brake. Generally, though, cold-formed members are made by running a coil of sheet steel through a series of rolling stands, each of which makes a small step in bending the sheet to its final form. But the equipment investment is still much less than that of the hot-rolled industry, and the end product coming out of the last roller stand can often be lifted by one person.

It is easy for a manufacturer of cold-formed steel sections to add a wrinkle here and there to try and get an edge on a competitor. For this reason, there is not much in the way of standardization of parts. Each manufacturer makes the section it thinks will best compete in the marketplace. They may be close but rarely exactly like products by other manufacturers used in similar applications.

The sections which appear in Part I of the AISI Cold-Formed Steel Manual (Ref. 1.2) are "commonly used" sections representing an average survey of suppliers, but they are not "standard" in the sense that every manufacturer makes these sizes.

1.4 HISTORY OF COLD-FORMED STANDARDS

In order to ensure that all designers and manufacturers of cold-formed steel products were competing fairly, and to provide guidance to building codes, some sort of national consensus standard was needed. Such a standard was first developed by the American Iron and Steel Institute (AISI) in 1946.

The first AISI Specification for the Design of Cold-Formed Steel Members was based largely on research work done under the direction of Professor George Winter at Cornell University. Research was done between 1939 and 1946 on beams, studs, roof decks, and connections under the supervision of an AISI Technical Subcommittee, which prepared the first edition of the specification and manual. The specification has been revised and updated as new research reveals better design methods. A complete chronology of all the editions follows:

First edition: 1946
Second edition: 1956
Third edition: 1960
Fourth edition: 1962
Fifth edition: 1968
Sixth edition: 1980
Seventh edition: 1986
Addendum: 1989
First edition LRFD: 1991
Combined ASD and LRFD and 50th anniversary
    edition: 1996
Addendum to the 1996 edition: 1999

Further details on the history can be found in Ref. 1.3.

The 1996 edition of the specification combines allowable stress design (ASD) and load and resistance factor design (LRFD) into one document. It was the feeling of the Specification Committee that there should be only one formula for calculating ultimate strength for various limit states. For example, the moment causing lateral-torsional buckling of a beam should not depend on whether one is using ASD or LRFD. Once the ultimate moment is determined, the user can then divide it by a factor of safety (Ω) and compare it to the applied moment, as in ASD, or multiply it by a resistance (ϕ) factor and compare it to an applied moment which has been multiplied by appropriate load factors, as in LRFD.

The equations in the 1996 edition are organized in such a way that any system of units can be used. Thus, the same equations work for either customary or SI units. This makes it one of the most versatile standards ever developed.

The main purpose of this book is to discuss the 1996 version of AISI's Specification for the Design of Cold-Formed Structural Steel Members, along with the 1999 Addendum, and to demonstrate with design examples how it can be used for the design of cold-formed steel members and frames.

Other international standards for the design of cold-formed steel structures are the Australia/New Zealand Standard AS/NZS 4600 (Ref. 1.4), which is based mainly on the 1996 AISI Specification with some extensions for high-strength steels, the British Standard BS 5950-Part 5 (Ref. 1.5), the Canadian Standard CAN/CSA S136 (Ref 1.6) and the Eurocode 3 Part 1.3 (Ref. 1.7), which is still a prestandard. All these international standards are only in limit states format.

1.5 COMMON SECTION PROFILES AND APPLICATIONS OF COLD-FORMED STEEL

Cold-formed steel structural members are normally used in the following applications.

Roof and wall systems of industrial, commercial and agricultural buildings: Typical sections for use in roof and wall systems are Z- (zee) or C- (channel) sections, used as purlins and girts, or sometimes beams and columns. Typically, formed steel sheathing or decking spans across these members and is fastened to them with self-drilling screws through the "valley" part of the deck. In most cases, glass fiber insulation is sandwiched between the deck and the purlins or girts Concealed fasteners can also be used to eliminate penetrations in the sheathing. Typical purlin

5

sections and profiles are shown in Figure 1.1, and typical deck profiles as shown in Figure 1.2. Typical fixed clip and sliding clip fasteners are shown in Figure 1.3.

Steel racks for supporting storage pallets: The uprights are usually channels with or without additional rear flanges, or tubular sections. Tubular or pseudotubular sections such as lipped channels intermittently welded toe-to-toe are normally used as pallet beams. Typical sections are shown in Figure 1.4, and a complete steel storage rack in Figure 1.5. In the United States the braces are usually welded to the uprights, whereas in Europe, the braces are normally bolted to the uprights, as shown in Figure 1.4a.

Structural members for plane and space trusses: Typical members are circular, square, or rectangular hollow sections both as chords and webs, usually with welded joints as shown in Figure 1.6a. Bolted joints can also be achieved by bolting onto splice plates welded to the tubular sections. Channel section chord members can also be used with tubular braces bolted or welded into the open sections as shown in Figure 1.6b. Cold-formed channel and Z sections are commonly used for the chord members of roof trusses of steel-framed housing. Trusses can also be fabricated from cold-formed angles.

Frameless stressed-skin structures: Corrugated sheets or sheeting profiles with stiffened edges are used to form small structures up to a 30-ft clear span with no interior

Z (Zee) sections   C (Channel) section

FIGURE 1.1   Purlin sections.

6

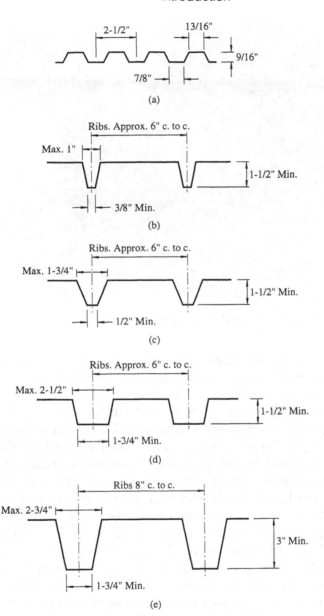

FIGURE 1.2 Deck profiles: (a) form deck (representative); (b) narrow rib deck type NR; (c) intermediate rib deck type IR; (d) wide rib deck type WR; (e) deep rib deck type 3DR.

FIGURE 1.3   Sliding clip and fixed clip.

framework. Farm buildings, storage sheds, and grain bins are typical applications, although such construction has historically been used for industrial buildings as well.

Residential framing: Lipped and unlipped channels, made to the same dimensions as nominal 2×4s, and sometimes known as "steel lumber," are typically used in the walls of residential buildings. Larger channel sections are used as floor and ceiling joists, and roof trusses are commonly made of small channel sections screwed or bolted together. Some examples are shown in Figure 1.7b for a simple roof truss and in Figure 1.7a for wall framing.

Steel floor and roof deck: Formed steel deck is laid across steel beams to provide a safe working platform and a form for concrete. It is normally designated as a wide rib, intermediate rib, or narrow rib deck. Some deck types have a flat sheet attached to the bottom of the ribs which creates hollow cells providing raceways for electrical and other cables. This bottom sheet may be perforated for an acous-

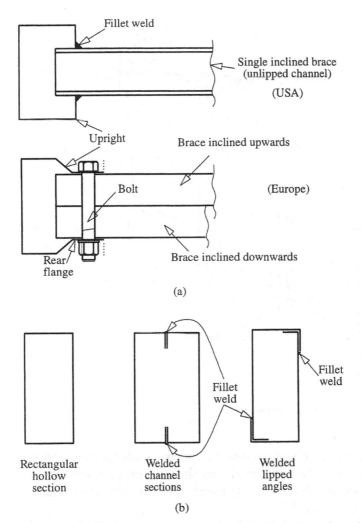

FIGURE 1.4   Storage rack sections: (a) plan view of uprights and bracing; (b) pallet beam sections.

tical ceiling. Typical examples of each of these decks are shown in Figure 1.2.

Some decks have embossments in the sloping sides of the rib that engage the concrete slab as a kind of shear key and permit the deck to act compositely with the concrete.

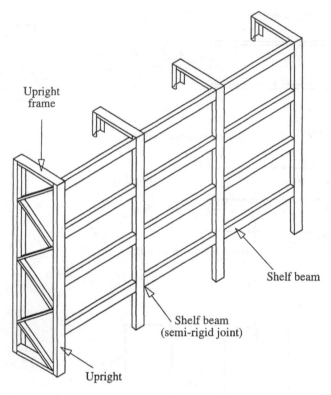

Upright
frame

Shelf beam

Shelf beam
(semi-rigid joint)

Upright

FIGURE 1.5   Steel storage rack.

It is not common to use a concrete slab on roofs, so roof deck is generally narrow rib so that it can provide support for insulating board. A typical narrow rib roof deck is shown in Figure 1.2.

Utility poles: All of the foregoing may lead the reader to think that "cold-formed" means "thin sheet." It is often surprising to one not familiar with the subject that cold-formed structures may be fabricated from plates up to 1 in. thick. (In fact the 1996 AISI Specification declares that it covers all steel up to an inch thick.) A good example is a type of vertical tapered pole that may be used for street lights or traffic signals. It is difficult to make a 90° bend in a

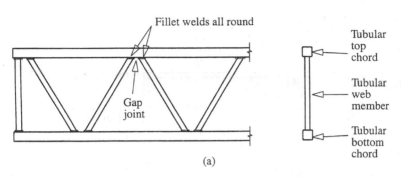

(a)

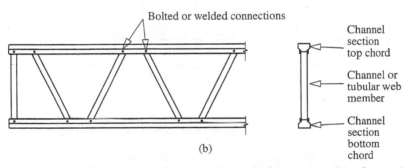

(b)

FIGURE 1.6 Plane truss frames: (a) tubular truss; (b) channel section truss.

1 in.-thick plate, and even if it could be done it would probably crack the plate, but it is feasible to make a 60° bend. Such poles are generally octagonal in cross section and are made in halves that are then welded together. Because they are tapered, they can be stacked to attain greater height. An example of this kind of cold-formed structure can be seen in Figure 1.8.

Automotive applications: All major structural elements can be used, but normally hat sections or box sections are used. Several publications on this subject are available from the American Iron and Steel Institute.

Grain storage bins or silos: Grain bins usually consist of curved corrugated sheets stiffened by hat or channel

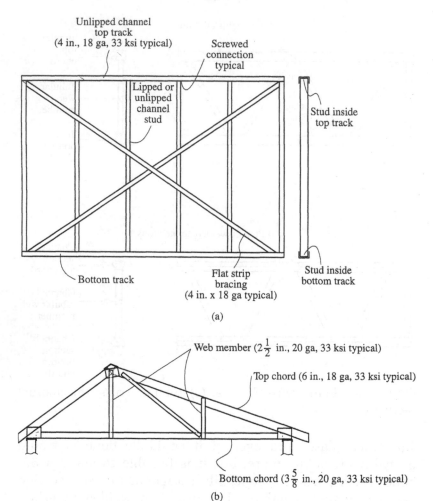

Unlipped channel
top track
(4 in., 18 ga, 33 ksi typical)

Screwed
connection
typical

Lipped or
unlipped
channel
stud

Stud inside
top track

Bottom track

Flat strip
bracing
(4 in. x 18 ga typical)

Stud inside
bottom track

(a)

Web member ($2\frac{1}{2}$ in., 20 ga, 33 ksi typical)

Top chord (6 in., 18 ga, 33 ksi typical)

Bottom chord ($3\frac{5}{8}$ in., 20 ga, 33 ksi typical)

(b)

FIGURE 1.7   Residual construction: (a) wall framing; (b) roof truss.

sections. While the AISI Specification for cold-formed structures does not deal specifically with such structures, the
same principles can be applied. A good reference on this
kind of structure is Gaylord and Gaylord (Ref. 1.8).

Cold-formed tubular members: Hollow structural
sections (HSS) may be made by cold-roll-forming to produce

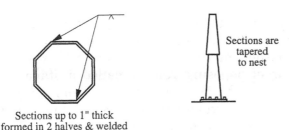

Sections are
tapered
to nest

Sections up to 1" thick
formed in 2 halves & welded

FIGURE 1.8   Cold-formed steel utility poles.

a round, which is then closed by electric resistance welding (ERW). The round shape can then be used as is or further formed into a square or rectangle. Examples are shown in Figure 1.9.

1.6 MANUFACTURING PROCESSES

Cold-formed members are usually manufactured by one of two processes: (1) roll forming and (2) brake forming.

1.6.1 Roll Forming

Roll forming consists of feeding a continuous steel strip through a series of opposing rolls to progressively deform the steel plastically to form the desired shape. Each pair of rolls produces a fixed amount of deformation in a sequence

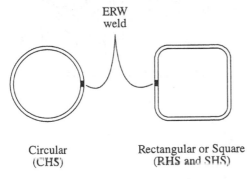

ERW
weld

Circular
(CHS)

Rectangular or Square
(RHS and SHS)

FIGURE 1.9   Typical tubular sections.

13

of the type shown in Figure 1.10. In this example, a Z section is formed by first developing the bends to form the lip stiffeners and then producing the bends to form the flanges. Each pair of opposing rolls is called a stage, as shown in Figure 1.11a. In general, the more complex the cross-sectional shape, the greater the number of stages that are required. In the case of cold-formed rectangular hollow sections, the rolls initially form the section into a circular section and a weld is applied between the opposing edges of the strip before final rolling (called sizing) into a square or rectangle.

1.6.2 Brake Forming

Brake forming involves producing one complete fold at a time along the full length of the section, using a machine called a press brake, such as the one shown in Figure 1.11b. For sections with several folds, it is necessary to move the steel plate in the press and to repeat the braking operation several times. The completed section is then removed from the press and a new piece of plate is inserted for manufacture of the next section.

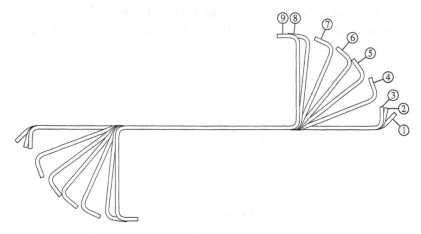

FIGURE 1.10   Typical roll-forming sequence for a Z-section.

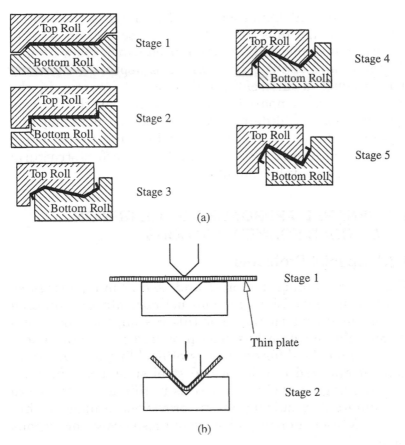

FIGURE 1.11   Cold-forming tools: (a) roll-forming tools; (b) brake press dies.

Roll forming is the more popular process for producing large quantities of a given shape. The initial tooling costs are high, but the subsequent labor content is low. Press braking is normally used for low-volume production where a variety of shapes are required and the roll-forming tooling costs cannot be justified. Press braking has the further limitation that it is difficult to produce continuous lengths exceeding about 20 ft.

A significant limitation of roll forming is the time it takes to change rolls for a different size section. Consequently, adjustable rolls are often used which allow a rapid change to a different section width or depth. Roll forming may produce a different set of residual stresses in the section when compared with braking, so the section strength may be different in cases where buckling and yielding interact. Also, corner radii tend to be much larger in roll-formed sections, and this can affect structural actions such as web crippling.

1.7 GENERAL APPROACH TO THE DESIGN OF COLD-FORMED SECTIONS

1.7.1 Special Problems

The use of thinner material and cold-forming processes result in special design problems not normally encountered in hot-rolled construction. For this reason, the AISI Cold-Formed Specification has been produced to give designers guidance on the different modes of buckling and deformation encountered in cold-formed steel structures. In addition, welding and bolting practices in thinner sections are also different, requiring design provisions unique to thin sheets. A brief summary of some of these design provisions follows.

1.7.2 Local Buckling and Post-Local Buckling of Thin Plate Elements

The thicknesses of individual plate elements of cold-formed sections are normally small compared to their widths, so local buckling may occur before section yielding. However, the presence of local buckling of an element does not necessarily mean that its load capacity has been reached. If such an element is stiffened by other elements on its edges, it possesses still greater strength, called "postbuckling strength." Local buckling is *expected* in most cold-formed sections and often ensures greater economy than a heavier section that does not buckle locally.

1.7.3 Effective Width Concept

In the first specification published by the American Iron and Steel Institute in 1946, steel designers were introduced to the concept of "effective width" of stiffened elements of a cold-formed section for the first time. The notion that a flat plate could buckle and still have strength left—postbuckling strength as it was called—was a new concept in steel specifications. To have this postbuckling strength required that the plate be supported along its edges or "stiffened" by some element that was attached at an angle, usually a right angle. These stiffening elements are achieved in cold-formed steel by bending the sheet.

Because cold-formed members typically have very high width-to-thickness ratios, they tend to buckle elastically under low compressive stress. However, the stiffened edges of the plate remain stable and a certain width of the plate close to the corners is still "effective" in resisting further compressive load. The problem is to determine how much of the original width of the plate is still effective. This is called "effective width," and formulas for calculating it were developed under the leadership of Dr. George Winter at Cornell University in the early 1940s. These effective width formulas appeared in the first Cold-Formed Specification in 1946 and remained unchanged until 1986.

Plates that had a stiffening element on only one edge were called "unstiffened." These did not require calculation of an effective width, but were designed on the basis of a reduced stress.

Until 1986, there were only *stiffened* and *unstiffened* elements. Examples of each are shown in Figure 1.12a. An element was deemed stiffened if it had an adequate stiffener on both edges of the element. The stiffener could be "edge" or "intermediate," as shown in Figure 1.12b. The

17

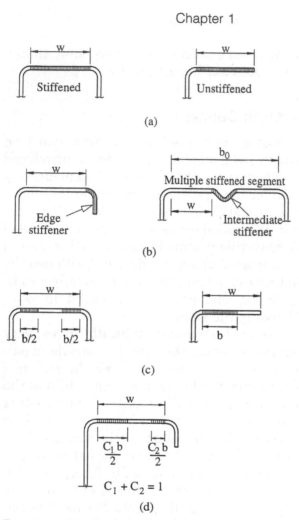

FIGURE 1.12 Compression elements: (a) compression elements; (b) stiffeners; (c) effective widths; (d) effective width for a partially stiffened element.

stiffener's adequacy was a clearly defined limiting moment of inertia, dependent on the slenderness—that is, the width-to-thickness ratio—of the element being stiffened. If stiffened elements were very slender, the real width might have to be reduced to an *effective width*, as shown in Figure 1.12c. The effective area thus computed for a complete section, when divided by the gross area, produced an area reduction factor called Q_a.

Effective area was not calculated for unstiffened elements. The lower buckling stress on an unstiffened element was calculated according to formulas which were a function of the slenderness of the element. The resulting stress, when divided by the design stress, usually $0.6F_y$, produced a stress reduction factor, called Q_s. The total reduction on a section in compression was $Q_a \times Q_s$.

In the 1986 edition of the specification, a major shift in philosophy was made. Now *all* compression elements are treated with an effective width approach, as shown in Figure 1.12c. There is *one basic effective width equation* and the only difference that separates one element from another is the plate buckling constant, k, which is discussed in detail in Chapter 4. Even though the specification still speaks of stiffened and unstiffened elements, most elements are stiffened to some degree according to their edge conditions and stress gradients, and one can begin to think in terms of only one kind of element: a *partially stiffened* element, as shown in Figure 1.12d. Q_a and Q_s are no more.

Formerly, in a section such as a channel that has a flange stiffened by a web on one side and a lip on the other, the flange was considered a stiffened element. Now, there is a distinction between the flange of a channel and the flange of a hat section which is attached to webs on both sides. The channel flange is now called an *edge-stiffened* element, while the flange of the hat is still called a *stiffened* element. The edge stiffener usually produces effective widths distributed as shown in Figure 1.12d.

The web of either type of section, which has a portion of its depth in compression, is also now treated with an effective width approach, as are all elements with a stress gradient. The only items that change are the plate buckling coefficient (k) and the distribution of the effective widths, as discussed in Chapter 4.

1.7.4 Propensity for Twisting

Cold-formed sections are normally thin, and consequently they have a low torsional stiffness. Many of the sections produced by cold-forming are singly symmetric with their shear centers eccentric from their centroids, as shown in Figure 1.13a. Since the shear center of a thin-walled beam is the axis through which it must be loaded to produce flexural deformation without twisting, any eccentricity of the load from this axis generally produces considerable torsional deformations in a thin-walled beam, as shown in Figure 1.13a. Consequently, beams usually require torsional restraints at intervals or continuously along them to prevent torsional deformations. Methods for bracing channel and Z sections against torsional and lateral deformations are found in Section D of the 1996 AISI Specification and in Chapter 10 of this book.

For a column axially loaded along its centroidal axis, the eccentricity of the load from the shear center axis may cause buckling in the torsional-flexural mode as shown in Figure 1.13b at a lower load than the flexural buckling mode, also shown in Figure 1.13b. Hence, the designer must check for the torsional-flexural mode of buckling using methods described in Chapter 7 of this book and in Section C4 of the AISI Specification.

Beams such as channel and Z purlins and girts may undergo lateral-torsional buckling because of their low torsional stiffness. Hence, design equations for lateral-torsional buckling of purlins with different bracing conditions are given in Section C3 of the AISI Specification and are described in Chapter 5 of this book.

1.7.5 Distortional Buckling

Sections which are braced against lateral or torsional-flexural buckling may undergo a mode of buckling commonly known as distortional buckling, as shown in Figure 1.14. This mode can occur for members in flexure

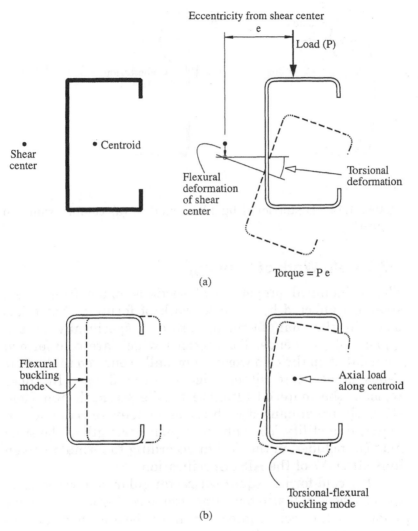

FIGURE 1.13 Torsional deformations: (a) eccentrically loaded channel beam; (b) axially loaded channel column.

or compression. In the 1996 edition of the AISI Specification, distortional buckling of edge-stiffened elements is partially accounted for by Section B4.2. Further research is underway to include specific design rules.

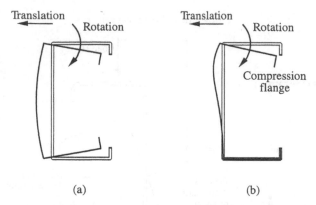

(a)   (b)

FIGURE 1.14 Distortional buckling modes: (a) compression; (b) flexure.

1.7.6 Cold Work of Forming

The mechanical properties of sections made from sheet steel are affected by the cold work of forming that takes place in the manufacturing process, specifically in the regions of the bends. For sections which are cold-formed from flat strip the cold work is normally confined to the four bends adjacent to either edge of each flange. In these regions, the material ultimate tensile strength and yield strength are enhanced with a commensurate reduction in material ductility. The enhanced yield strength of the steel may be included in the design according to formulae given in Section A7 of the AISI Specification.

For cold-formed square or rectangular hollow sections, the flat faces will also have undergone cold work as a result of forming the section into a circular tube and then reworking it into a rectangle or square. In this case, it is very difficult to compute theoretically the enhancement of yield strength in the flats, so the AISI Specification allows the measured yield strength of the steel after forming to be used in design where the yield strength is determined according to the procedures described in ASTM A500. The distribution of ultimate tensile strength and yield strength measured in a cold-formed square hollow section is shown in Figure 1.15. The distribution indicates that the properties are reasonably uniform across the flats (except at the weld location) with a yield strength of approximately 57 ksi for a nominal yield stress of 50 ksi. This is substantially higher than the yield strength of the plate before forming, which is normally approximately 42 ksi. The enhancement of the yield stress in the corners is very substantial, with an average value of approximately 70 ksi.

1.7.7 Web Crippling Under Bearing

Web crippling at points of concentrated load and supports can be a critical problem in cold-formed steel structural members and sheeting for the following reasons:

1. In cold-formed design, it is often not practical to provide load bearing and end bearing stiffeners.

Distribution of yield stress (×) and ultimate tensile strength (○)

FIGURE 1.15 Tensile properties in a cold-formed square hollow section (Grade C to ASTM A500).

This is always the case in continuous sheeting and decking spanning several support points.

2. The depth-to-thickness ratios of the webs of cold-formed members are usually larger than for hot-rolled structural members.
3. In many cases, the webs are inclined rather than vertical.
4. The load is usually applied to the flange, which causes the load to be eccentric to the web and causes initial bending in the web even before crippling takes place. The larger the corner radius, the more severe is the case of web crippling.

Section C3.4 of the AISI Specification provides design equations for web crippling of beams with one- and two-flange loading, stiffened or unstiffened flanges, and sections having multiple webs.

At interior supports, beams may be subjected to a combination of web crippling and bending. Section C3.5 provides equations for this condition. Likewise, such sections may be subjected to a combination of shear and bending. Section C3.3 covers this condition.

1.7.8 Connections

A. Welded Connections

In hot-rolled steel fabrication, the most common welds are fillet welds and partial or complete penetration groove welds. There is a much wider variety of welds available in cold-formed construction, such as arc spot welds (sometimes referred to as puddle welds), flare bevel groove welds, resistance welds, etc. In hot-rolled welding, generally the two pieces being joined are close to the same thickness, so it is easy for the welder to set the amperage on his welding machine. In cold-formed steel, there is often a great difference in the thicknesses of the pieces being joined. For example, welding a 22 gauge deck to a $\frac{3}{4}$-in. beam flange requires completely different welding skills. To get enough heat to fuse the thick flange means that the thin sheet may be burned through and if there is not intimate contact between the two pieces, a poor weld will result. In these cases, weld washers, thicker than the sheet, may be used to improve weld quality.

Another difference between hot-rolled welding and light-gauge welding is the presence of round corners. In welding two channels back-to-back, for example, there is a natural groove made by the corner radii and this can be used as a repository for weld metal. Such welds are called flare bevel groove welds.

Tests of welded sheet steel connections have shown that failure generally occurs in the material around the weld rather than in the weld itself. Provisions for welded connections are found in Section E2 of the AISI Specification and are discussed in Chapter 9.

B. Bolted Connections

Bolted connections in cold-formed construction differ primarily in the ratio of the bolt diameter to the thickness of the parts being joined. Thus, most bolted connections are controlled by bearing on the sheet rather than by bolt shear. The "net section fracture" that is usually checked in heavy plate material rarely occurs. Provisions for bolted connections appear in Section E3 and are discussed in Chapter 9.

C. Screw Connections

Self-drilling screws, self-tapping screws, sheet metal screws, and blind rivets are all devices for joining sheet to sheet or sheet to heavier supporting material. Of these, the self-drilling screw is the most common for structural applications. It is used for securing roof and wall sheets to purlins and girts and making connections in steel framing for residential or commercial construction. Sheet metal screws and rivets are mostly used for non-structural applications such as flashing, trim, and accessories. Pullover and pullout values for screws may be calculated with the equations in Section E4 or values may be obtained from the manufacturer. The design of screw connections is discussed in Chapter 9.

D. Clinching

A relatively new method for joining sheets is called clinching. No external fastener is used here; a special punch deforms the two sheets in such a way that they are held together. At this time, no provisions for determining the capacity of this type of connection exist in the AISI Specification, so the testing provisions of Section F of the AISI Specification should be used.

1.7.9 Corrosion Protection

The main factor governing the corrosion resistance of cold-formed steel sections is not the base metal thickness but the type of protective treatment applied to the steel. Cold-formed steel has the advantage of applying protective coatings to the coil before roll forming. Galvanizing is a process of applying a zinc coating to the sheet for corrosion protection and is purely functional. A paint finish on steel can serve as corrosion protection *and* provide an attractive finish for exterior panels. A painted or galvanized coil can be passed through the rolls and the finish is not damaged. A manufacturer of exterior panels can now order coils of steel in a variety of colors and the technology of the factory finish has improved to the point that the paint will not crack when a bend is made in the sheet.

Other than paint, the most common types of coatings for sheet steel are zinc (galvanizing), aluminum, and a combination of zinc and aluminum. These are all applied in a continuous coil line operation by passing the strip through a molten metal bath followed by gas wiping to control the amount of coating applied. Typical coating designations are G60 and G90 for galvanized product and T2 65 and T2 100 for aluminum-coated sheet. The numbers in these designations refer to the total coating weight on *both sides* of the sheet in hundredths of an ounce per square foot (oz/ft^2) of sheet area. For example, G60 means a minimum of 0.60 ounce of zinc per square foot. This corresponds to a 1-mil thickness on each side. A G90 coating has 0.90 ounce per square foot and a thickness of 1.5 mils (Ref. 1.9).

1.7.10 Inelastic Reserve Capacity

Generally, the proportions of cold-formed sections are such that the development of a fully plastic cross-section is unlikely. However, there is a provision in Section C3.1.1(b), Procedure II, that permits the use of inelastic reserve capacity for stockier sections, such as those found in rack structures. Here, compressive strains up to three times the yield strain are permitted for sections satisfying certain slenderness limits. In this case, the design moment may not exceed the yield moment by more than 25%.

1.8 TWO DESIGN METHODS

The 1996 AISI Specification for Cold-Formed Steel Members is a unique document in that it simultaneously satisfies those wishing to use the time-tested allowable stress design (ASD) method and those wishing to use the more modern load and resistance factor design (LRFD) method. At the same time, all the equations in the specification are formulated in such terms that any system of units may be used. Thus, four permutations are available in one reference:

> ASD with customary or English units
> ASD with SI units
> LRFD with customary or English units
> LRFD with SI units

Probably 90% of the 1996 specification deals with finding failure loads or nominal capacities. The process is the same whether one is using ASD or LRFD. The decision as to which method to use is only made at the end, when the nominal capacity is either divided by a safety factor (Ω) or multiplied by a resistance factor (ϕ). Of course, in ASD the capacity is compared to the service loads or forces and in LRFD it is compared to factored forces.

1.8.1 Allowable Stress Design

The basis for ASD is the use of fractions of the material's yield stress as a limiting stress. Put another way, the yield stress is divided by a *safety factor* to determine the maximum stress allowed for a particular structural action. For example, the allowable bending stress may be 0.6 × yield stress. The inverse of this coefficient, 1.67, is Ω. Each type of stress, tension, compression, bending, shear, etc., has its own corresponding *allowable* stress which cannot be exceeded by the *actual* stresses. Ever since engineers have been calculating the forces and stresses in structures, this has been the method used. The only changes have been in the *way* the allowable stress is determined, and this is the function of a design standard such as AISC or AISI specifications.

1.8.2 Load and Resistance Factor Design

The philosophy of LRFD is that all the variabilities that the old safety factor was supposed to take into account should be divided into two factors: one representing the load variability and the other accounting for all those things that go into making variability in resistance, such as dimensional tolerances, material variations, errors in fabrication, errors in erection, analytical assumptions, and others. These are called, respectively, the *load factor* and *resistance factor*. Typical frequency distribution curves for load effects and corresponding resistances are shown in Figure 1.16.

In addition, the load factor is not constant for all types of loading. The load with least variability, such as dead load, can have a smaller factor. This is an advantage for structures with heavy dead load, one that is not available in ASD. In most cold-formed steel structures, the dead load tends to be a small percentage of the total gravity load, thus partially negating the advantages of LRFD. In order to produce designs that are consistent with the more traditional ASD, the LRFD specification was "calibrated" at a live/dead load ratio of 5/1.

1.9 LOAD COMBINATIONS

The load combination in Section A5.1.2 for allowable stress design and in Section A6.1.2 for load and resistance factor design follow the standard ASCE-7-98 (Ref. 1.10), but with a few important exceptions. The traditional 1/3 stress increase is no longer permitted in ASCE-7 for ASD, but a 0.75 reduction factor may be used when wind or earthquake loads are used in combination with other live loads.

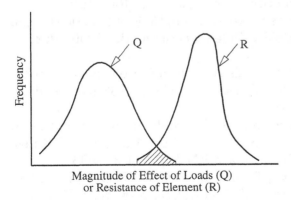

FIGURE 1.16   Frequency distribution curves.

One application that is not found in hot-rolled steel is that of a steel deck which supports a concrete floor. A special load combination for erection purposes, including the weight of the wet concrete as a live load, appears in the commentary to A6.1.2 in the AISI specification. Another difference involves a reduced wind load factor for secondary members, that is, sheeting, purlins, girts and other members. These members can be designed with a lower reliability index than primary structural members of long-standing practice.

REFERENCES

1.1 American Institute of Steel Construction, Load and Resistance Factor Design Specification for Structural Steel Buildings, 1993.

1.2 American Iron and Steel Institute, Cold-Formed Steel Design Manual, 1996.

1.3 Yu, W-W, Wolford, D. S., and Johnson, A. L., Golden Anniversary of the AISI Specification, Center for Cold-Formed Steel Structures, University of Missouri-Rolla, 1996.

1.4 Standards Australia/Standards New Zealand, Cold-Formed Steel Structures, AS/NZS 4600:1996.

1.5 British Standards Institution, Code of Practice for the Design of Cold-Formed Sections, BS 5950, Part 5, 1986.

1.6 Canadian Standards Association, Cold-Formed Steel Structural Members, CAN/CSA S136-94, Rexdale, Ontario, 1994.

1.7 Comité Européen de Normalisation, Eurocode 3: Design of Steel Structures, Part 1.3: General Rules, European Prestandard ENV 1993-1-3, 1996.

1.8 Gaylord, E. H., Jr, and Gaylord, C. N., Design of Steel Bins for Storage of Bulk Solids, New York, Prentice-Hall, 1984.

1.9 Bethlehem Steel Corporation, Technical Bulletin, Coating Weight and Thickness Designations for Coated Sheet Steels, March 1995.

1.10 American Society of Civil Engineers, Minimum Design Loads for Buildings and Other Structures, 1998.

2

Materials and Cold Work of Forming

2.1 STEEL STANDARDS

The AISI Specification allows the use of steel to the following specifications:

ASTM A36/A36M,   Carbon structural steel

ASTM A242/A242M,   High-strength low-alloy (HSLA) structural steel

ASTM A283/A283M,   Low- and intermediate-tensile-strength carbon steel plates

ASTM A500,   Cold-formed welded and seamless carbon steel structural tubing in rounds and shapes

ASTM A529/A529M,   High-strength carbon-manganese steel of structural quality

ASTM A570/A570M,   Steel, sheet and strip, carbon, hot-rolled, structural quality

ASTM A572/A572M,   High strength low alloy columbium-vanadium structural steel

ASTM A588/A588M, High-strength low-alloy structural steel with 50 ksi (345 MPa) minimum yield point to 4 in. (100 mm) thick

ASTM A606, Steel, sheet and strip, high-strength, low-alloy, hot-rolled and cold-rolled, with improved atmospheric corrosion resistance

ASTM A607, Steel, sheet and strip, high-strength, low-alloy, columbium or vanadium, or both, hot-rolled and cold-rolled

ASTM A611 (Grades A, B, C, and D), Structural steel (SS), sheet, carbon, cold-rolled

ASTM A653/A653M (SS Grades 33, 37, 40, and 50 Class 1 and Class 3; HSLAS types A and B, Grades 50, 60, 70 and 80), steel sheet, zinc-coated (galvanized) or zinc-iron alloy-coated (galvannealed) by the hot-dip process

ASTM A715 (Grades 50, 60, 70, and 80) Steel sheet and strip, high-strength, low-alloy, hot-rolled, and steel sheet, cold rolled, high-strength, low-alloy with improved formability

ASTM A792/A792M (Grades 33, 37, 40, and 50A), Steel sheet, 55% aluminum-zinc alloy-coated by the hot-dip process

ASTM A847 Cold-formed welded and seamless high-strength, low-alloy structural tubing with improved atmospheric corrosion resistance

ASTM A875/875M (SS Grades 33, 37, 40, and 50 Class 1 and Class 3; HSLAS types A and B, Grades 50, 60, 70, and 80) steel sheet, zinc-5% aluminum alloy-coated by the hot-dip process

Tubular sections manufactured according to ASTM A500 have four grades of round tubing (A, B, C, D) and four grades (A, B, C, D) of shaped tubing. Details of stress grades are shown in Table 2.1.

Steels manufactured according to ASTM A570 cover hot-rolled carbon steel sheet and strip of structural quality

TABLE 2.1 Minimum Values of Yield Point, Tensile Strength, and Elongation

ASTM designation	Product	Grade	F_y (ksi) (min)	F_u (ksi) (min/max)	Permanent elongation in 2 in. (min)	F_u/F_y (min)
A500-98	Round tubing	A	33	45	25	1.36
		B	42	58	23	1.38
		C	46	62	21	1.35
		D	36	58	23	1.61
	Shaped tubing	A	39	45	25	1.15
		B	46	58	23	1.26
		C	50	62	21	1.24
		D	36	58	23	1.61
A570/ A570M-96	Sheet and strip	30	30	49	21	1.63
		33	33	52	18	1.58
		36	36	53	17	1.47
		40	40	55	15	1.38
		45	45	60	13	1.33
		50	50	65	11	1.30
A6C7-96	Sheet and strip	Class 1	Class 1	Class 1		
		45	45	60	HR 23, CR 22	1.33
		50	50	65	HR 20, CR 20	1.30
		55	55	70	HR 18, CR 18	1.27
		60	60	75	HR 16, CR 16	1.25
		65	65	80	HR 14, CR 15	1.23
		70	70	85	HR 12, CR 14	1.21

35

TABLE 2.1 Continued

ASTM designation	Product	Grade	F_y (ksi) (min)	F_u (ksi) (min/max)	Permanent elongation in 2 in. (min)	F_u/F_y (min)
		Class 2	*Class 2*	*Class 2*		
		45	45	55	HR 23, CR 22	1.22
		50	50	60	HR 20, CR 20	1.20
		55	55	65	HR18, CR 18	1.18
		60	60	70	HR 16, CR 16	1.17
		65	65	75	HR 14, CR 15	1.15
		70	70	80	HR 12, CR 14	1.14
A611-97	Sheet	A	25	42	26	1.68
		B	30	45	24	1.50
		C types 1 and 2	33	48	22	1.45
		D types 1 and 2	40	52	20	1.30
A653/A653M-97	Sheet (galvanized or galvannealed)	SS 33	33	45	20	1.36
		37	37	52	18	1.41
		40	40	55	16	1.38
		50 (Class 1)	50	65	12	1.30
		50 (Class 3)	50	70	12	1.40
A792/A792M-97	Sheet aluminium-zinc alloy-coated	33	33	45	20	1.36
		37	37	52	18	1.41
		40	40	55	16	1.38
		50A	50	65	12	1.30

in cut lengths or coils. It is commonly used for purlins in the metal building industry. Details of stress grades are shown in Table 2.1.

Steels manufactured to ASTM A607 cover high-strength low-alloy columbium, or vanadium hot-rolled sheet and strip, or cold-rolled sheet, and is intended for applications where greater strength and savings in weight are important. Details of stress grades are shown in Table 2.1. Class 2 offers improved weldability and more formability than Class 1.

Steels manufactured to ASTM A611 cover cold-rolled carbon steel sheet in cut lengths or coils. It includes five strength levels designated A, B, C, D, and E. Grades A, B, C, D have moderate ductility, whereas Grade E is a full hard product with minimum yield point 80 ksi (550 MPa) and no specified minimum elongation. Details of stress grades are shown in Table 2.1.

Steels manufactured to ASTM A653/A653M-95 cover steel sheet, zinc-coated (galvanized) or zinc-iron alloy coated (galvannealed) by the hot-dip process. Structural quality and HSLA grades are available. Details of structural quality stress grades are shown in Table 2.1. SQ Grade 80 is similar to ASTM A611 Grade E.

Steels manufactured to ASTM A792 cover aluminum-zinc alloy-coated steel sheet in coils and cut lengths coated by the hot-dip process. The aluminum-zinc alloy composition by weight is normally 55% aluminum, 1.6% silicon, and the balance zinc. Details of stress grades are shown in Table 2.1.

Steels manufactured to ASTM A875 cover zinc-aluminum alloy coated steels coated by the hot-dip process with 5% aluminum. SS Grade 80 is similar to ASTM A611 Grade E.

2.2 TYPICAL STRESS-STRAIN CURVES

The stress-strain curve for a sample of 0.06-in.-thick sheet steel is shown in Figure 2.1. The stress-strain curve is

37

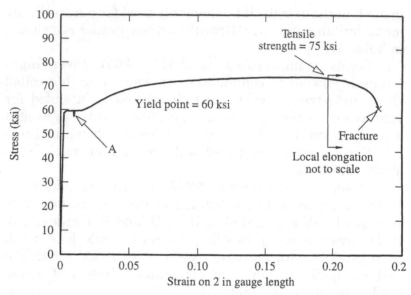

FIGURE 2.1 Stress-strain curve of a cold-rolled and annealed sheet steel.

typical of a cold-rolled and annealed low-carbon steel. A linear region is followed by a distinct plateau at 60 ksi, then strain hardening up to the ultimate tensile strength at 75 ksi. Stopping the testing machine for 1 min during the yield plateau, as shown at point A in Figure 2.1, allows relaxation to the static yield strength of 57.7 ksi. This reduction is a result of decreasing the strain rate to zero. The plateau value of 60 ksi is inflated by the effect of strain rate. The elongation on a 2-in. gauge length at fracture is approximately 23%. A typical measured stress-strain curve for a 0.06-in. cold-rolled sheet is shown in Figure 2.2. The steel has undergone cold-reducing (hard rolling) during the manufacturing process and therefore does not exhibit a yield point with a yield plateau as for the annealed steel in Figure 2.1. The initial slope of the stress-strain curve may be lowered as a result of the prework. The stress-strain curve deviates from linearity (proportional limit) at

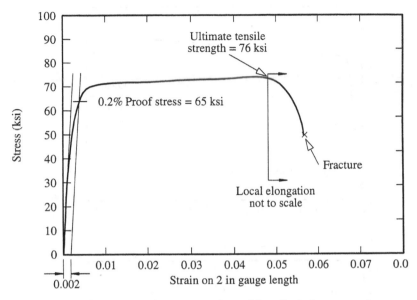

FIGURE 2.2 Stress-strain curve of a cold-rolled sheet steel.

approximately 36 ksi, and the yield point has been determined from 0.2% proof stress [as specified in AISI Manual
(Ref. 1.2)] to be 65 ksi. The ultimate tensile strength is
76 ksi, and the elongation at fracture on a 2-in. gauge
length is approximately 10%. The value of the elongation
at fracture shown on the graph is lower as a result of the
inability of an extensometer to measure the strain during
local elongation to the point of fracture without damage.
The graph in the falling region has been plotted for a
constant chart speed.

Typical stress-strain curves for a 0.08-in.-thick tube
steel are shown in Figure 2.3. The stress-strain curve in
Figure 2.3a is for a tensile specimen taken from the flats,
and the curve in Figure 2.3b is for a tensile specimen taken
from the corner of the section. The flat specimen displays a
yield plateau, probably as a result of strain aging (see
Section 2.4) following manufacture, whereas the corner

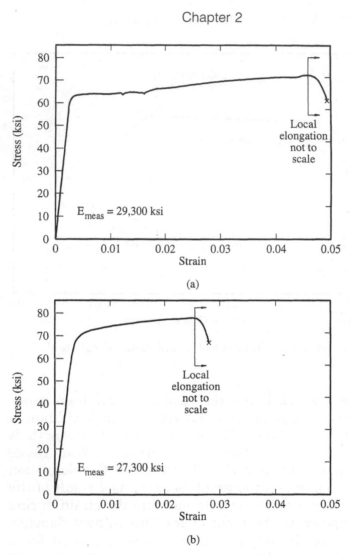

FIGURE 2.3   Stress-strain curves of a cold-formed tube: (a) stress-strain curve (flat); (b) stress-strain curve (corner).

specimen exhibits the characteristics of a cold worked steel. The Young's modulus (E_{meas}) of the corner specimen is lower than that of the flat specimen as a result of greater prework.

2.3 DUCTILITY

A large proportion of the steel used for cold-formed steel structures is of the type shown in Figures 2.2 and 2.3. These steels are high tensile and often have limited ductility as a result of the manufacturing processes. The question arises as to what is an adequate ductility when the steel is used in a structural member including further cold-working of the corners, perforations such as bolt holes, and welded connections. Two papers by Dhalla and Winter (Refs. 2.2, 2,3) attempt to define adequate ductility in this context.

Ductility is defined as the ability of a material to undergo sizable plastic deformation without fracture. It reduces the harmful effects of stress concentrations and permits cold-forming of a structural member without impairment of subsequent structural behavior. A conventional measure of ductility, according to ASTM A370, is the percent permanent elongation after fracture in a 2-in. gauge length of a standard tension coupon. For conventional hot-rolled and cold-rolled mild steels, this value is approximately 20–30%.

A tensile specimen before and after a simple tension test is shown in Figures 2.4a and 2.4b, respectively. After testing, the test length of approximately 3 in. has undergone a uniform elongation as a result of yielding and strain hardening. The uniform elongation is taken from the yield point up to the tensile strength as shown in Figure 2.4c. After the tensile strength has been reached, necking of the material occurs over a much shorter length (typically $\frac{1}{2}$ in. approximately) as shown in Figure 2.4b and ends when fracture of the test piece occurs. Elongation in the necking region is called local elongation. An alternative estimate of local ductility can be provided by calculating the ratio of the reduced area at the point of fracture to the original area. This measure has the advantage that it does not depend on a gauge length as does local elongation. However, it is more difficult to measure.

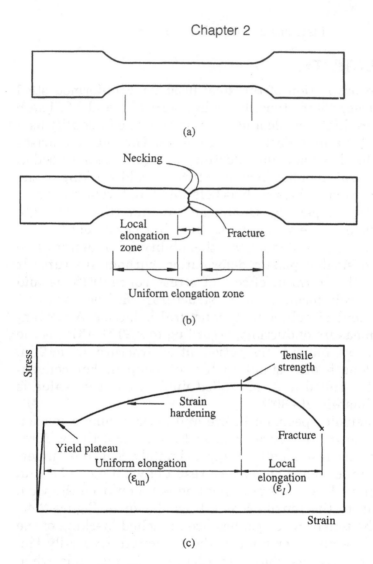

FIGURE 2.4 Ductility measurement: (a) tensile specimen before test; (b) tensile specimen after test; (c) stress-strain curve for hot formed steel.

Dhalla and Winter investigated a range of steels which had both large and small ratios of uniform elongation to local elongation. The purpose of the investigation was to determine whether uniform or local elongation was more beneficial in providing adequate ductility and to determine minimum values of uniform and local ductility. The steels tested ranged from cold-reduced steels, which had a uniform elongation of 0.2% and a total elongation of 5%, to annealed steels, which had a uniform elongation of 36% and a total elongation of 50%. In addition, a steel which had a total elongation of 1.3% was tested. Further, elastic-plastic analyses of steels containing perforations and notches were performed to determine the minimum uniform ductility necessary to ensure full yield of the perforated or notched section without the ultimate tensile strength being exceeded adjacent to the notch or perforation.

The following ductility criteria have been suggested to ensure satisfactory performance of thin steel members under essentially static load. They are included in Section A3.3 of the AISI Specification. The ductility criterion for *uniform elongation* outside the fracture is

$$\epsilon_{un} \geq 3.0\% \tag{2.1}$$

The ductility criterion for *local elongation* in a $\frac{1}{2}$ in. gauge length is

$$\epsilon_l \geq 20\% \tag{2.2}$$

The ductility criterion for the ratio of the tensile strength (F_u) to yield strength (F_y) is

$$\frac{F_u}{F_y} \geq 1.05 \tag{2.3}$$

In the AISI Specification, Section A3.3, 1.05 in Eq. (2.3) has been changed to 1.08.

The first criterion for uniform elongation is based mainly on elastic-plastic analyses. The second criterion for local elongation is based mainly on the tests of steels with low uniform elongation and containing perforations where it was observed that local elongations greater than 20% allowed complete plastification of critical cross sections. The third criterion is based on the observation that there is a strong correlation between uniform elongation and the ratio of the ultimate tensile strength to yield strength of a steel (F_u/F_y).

The requirements expressed by Eqs. (2.1) and (2.2) can be transformed into the conventional requirement specified in ASTM A370 for total elongation at fracture. Using a 2-in. gauge length and assuming necking occurs over a $\frac{1}{2}$ in. length, we obtain

$$\epsilon_{2\,\text{in.}} = \epsilon_{\text{un}} + \frac{0.5(\epsilon_l - \epsilon_{\text{un}})}{2.0} \qquad (2.4)$$

Substitution of Eqs. (2.1) and (2.2) in Eq. (2.4) produces

$$\epsilon_{2\,\text{in.}} = 7.25\% \qquad (2.5)$$

This value can be compared with those specified on a 2-in. gauge length in ASTM A570, which is 11% for a Grade 50 steel, and ASTM 607, which is 12% for a Grade 70 steel, as set out in Table 2.1.

Section A3.3.2 of the AISI Specification states that steels conforming to ASTM A653 SS Grade 80, A611 Grade E, A792 Grade 80, and A875 SS Grade 80, and other steels which do not comply with the Dhalla and Winter requirements may be used for particular multiple web configurations provided the yield point (F_y) used for design is taken as 75% of the specified minimum or 60 ksi, whichever is less, and the tensile strength (F_u) used for design is 75% of the specified minimum or 62 ksi, whichever is less.

A recent study of G550 steel to AS 1397 (Ref. 2.4) in 0.42-mm (0.016 in.) and 0.60-mm (0.024 in.) thicknesses has been performed by Rogers and Hancock (Ref. 2.5) to ascertain the ductility of G550 steel and to investigate the validity of Section A3.3.2. This steel is very similar to ASTM A653 SS Grade 80, A611 Grade E, A792 Grade 80, and A875 SS Grade 80. Three representative elongation distributions are shown in Figure 2.5 for unperforated

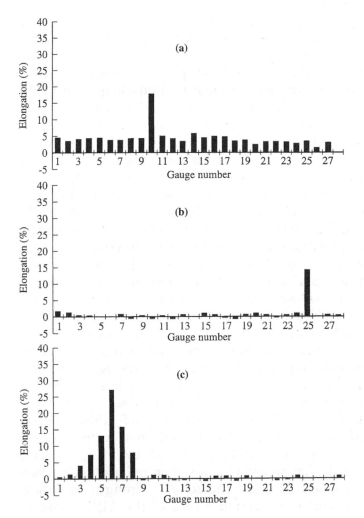

FIGURE 2.5 Elongation of 0.42-mm (0.016 in.) G550 (Grade 80) steel: (a) longitudinal specimen; (b) transverse specimen; (c) diagonal specimen.

tensile coupons taken from 0.42-mm (0.016 in.) G550 steel in (a) the longitudinal rolling direction of the steel strip, (b) the transverse direction, and (c) the diagonal direction. The coupons were each taken to fracture. In general, G550 coated longitudinal specimens have constant uniform elongation for each 0.10-in. gauge with an increase in percent elongation at the gauge in which fracture occurs as shown in Figure 2.5a. Transverse G550 specimens show almost no uniform elongation, but do have limited elongation at the fracture as shown in Figure 2.5b. The diagonal test specimen results in Figure 2.5c indicate that uniform elongation is limited outside of a $\frac{1}{2}$ in. zone around the fracture, while local elongation occurs in the fracture gauge as well as in the adjoining gauges. These results show that the G550 steel ductility depends on the direction from which the tensile coupons are obtained. The steel does not meet the Dhalla and Winter requirements described above except for uniform elongation in the longitudinal direction.

The study by Rogers and Hancock (Ref. 2.5) also investigated perforated tensile coupons to determine whether the net section fracture capacity could be developed at perforated sections. Coupons were taken in the longitudinal, transverse, and diagonal directions as described in the preceding paragraph. Circular, square, and diamond-shaped perforations of varying sizes were placed in the coupons. It was found that despite the low values of elongation measured for this steel, as shown in Figure 2.5, the load carrying capacity of G550 steel as measured in concentrically loaded perforated tensile coupons can be adequately predicted by using existing limit states design procedures based on net section fracture without the need to limit the tensile strength to 75% of 80 ksi as specified in Section A3.3.2 of the AISI Specification.

2.4 EFFECTS OF COLD WORK ON STRUCTURAL STEELS

Chajes, Britvec, and Winter (Ref. 2.6) describes a detailed study of the effects of cold-straining on the stress-strain characteristics of various mild carbon structural sheet steels. The study included tension and compression tests of cold-stretched material both in the direction of prior stretching and transverse to it. The material studied included

1. Cold-reduced annealed temper-rolled killed sheet coil
2. Cold-reduced annealed temper-rolled rimmed sheet coil
3. Hot-rolled semikilled sheet coil
4. Hot-rolled rimmed sheet coil

The terms rimmed, killed, and semikilled describe the method of elimination or reduction of the oxygen from the molten steel. In rimmed steels, the oxygen combines with the carbon during solidification and the resultant gas rises through the liquid steel so that the resultant ingot has a rimmed zone which is relatively purer than the center of the ingot. Killed steels are deoxidized by the addition of silicon or aluminum so that no gas is involved and they have more uniform material properties. Most modern steels are continuously cast and are aluminum or silicon killed.

A series of significant conclusions were derived in Ref. 2.6 and can be explained by reference to Figure 2.6. The major conclusions are the following:

1. Cold work has a pronounced effect on the mechanical properties of the material both in the direction of stretching and in the direction normal to it.
2. Increases in the yield strength and ultimate tensile strength as well as decreases in the ductility were found to be directly dependent upon the amount of cold work. This can be seen in Figure 2.6a where the curve for immediate reloading returns to the virgin stress-strain curve in the strain-hardening region.
3. A comparison of the yield strength in tension with that in compression for specimens taken both

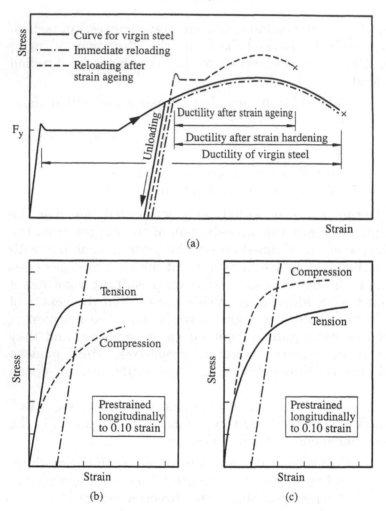

FIGURE 2.6 Effects of cold work on stress-strain characteristics: (a) effect of strain hardening and strain aging; (b) longitudinal specimen; (c) transverse specimen.

transversely and longitudinally demonstrates the Bauschinger effect (Ref. 2.7). In Figure 2.6b, longitudinal specimens demonstrate a higher yield in tension than compression, whereas in Figure 2.6c, transverse specimens of the same steel with the same degree of cold work demonstrate a higher yield in compression than in tension.

4. Generally, the larger the ratio of the ultimate tensile strength to the yield strength, the larger is the effect of strain hardening during cold work.

5. Aging of a steel occurs if it is held at ambient temperature for several weeks or for a much shorter period at a higher temperature. As demonstrated in Figure 2.6a, the effect of aging on a cold-worked steel is to

   a. Increase the yield strength and the ultimate tensile strength
   b. Decrease the ductility of the steel
   c. Restore or partially restore the sharp yielding characteristic

Highly worked steels, such as the corners of cold-formed tubes which have characteristics such as shown in Figure 2.3b, do not return to a sharp yielding characteristic. Some steels, such as the cold-reduced killed steel described in Ref. 2.6, may not demonstrate aging.

2.5 CORNER PROPERTIES OF COLD-FORMED SECTIONS

As a consequence of the characteristics of cold-worked steel described in Section 2.4, the forming process of cold-formed sections will result in an enhancement of the yield strength and ultimate tensile strength in the corners (bends) as demonstrated in Figure 1.15 for a square hollow section. For sections undergoing cold-forming as demonstrated in Figure 1.10, Section A7.2 of the AISI Specification provides formulae to calculate the enhanced yield strength (F_{yc}) of the corners. These formulae were derived from Ref. 2.6. A brief summary of these formulae and their theoretical and experimental bases follows.

The effective stress–effective strain (σ-ϵ) characteristic of the steel in the plastic part of the stress-strain curve is assumed to be described by the power function

$$\sigma = k\epsilon^n \tag{2.6}$$

where k is the strength coefficient and n is the strain-hardening exponent. The effective stress (σ) is defined by the Von Mises distortion energy yield criterion (Ref. 2.8), and the effective strain (ϵ) is similarly defined. In Ref. 2.9, k and n are derived from tests on steels of the types described in Section 2.4 to be

$$k = 2.80F_{yv} - 1.55F_{uv} \tag{2.7}$$

$$n = 0.255\left(\frac{F_{uv}}{F_{yv}}\right) - 0.120 \tag{2.8}$$

where F_{yv} is the virgin yield point and F_{uv} is the virgin ultimate tensile strength.

The equations for the analytical corner models were derived with the following assumptions:

1. The strain in the longitudinal direction during cold-forming is negligible, and hence a condition of plane strain exists.
2. The strain in the circumferential direction is equal in magnitude and opposite in sign to that in the radial direction on the basis of the volume constancy concept for large plastic deformations.
3. The effective stress-effective strain curve given by Eq. (2.6) can be integrated over the area of a strained corner.

The resulting equation for the tensile yield strength of the corner is

$$F_{yc} = \frac{B_c F_{yv}}{(R/t)^m} \tag{2.9}$$

where

$$B_c = \frac{kb}{F_{yv}} \tag{2.10}$$

$$b = 0.045 - 1.315n \tag{2.11}$$

$$m = 0.803n \tag{2.12}$$

and R is the inside corner radius. Equations (2.11) and (2.12) were derived empirically from tests.

Substituting Eqs. (2.7) and (2.8) into Eqs. (2.10) and (2.11) and eliminating b produce

$$B_c = 3.69\left(\frac{F_{uv}}{F_{yv}}\right) - 0.819\left(\frac{F_{uv}}{F_{yv}}\right)^2 - 1.79 \tag{2.13}$$

which is Eq. A7.2-3 in the AISI Specification. Substituting Eq. (2.8) into Eq. (2.12) produces

$$m = 0.192\left(\frac{F_{uv}}{F_{yv}}\right) - 0.068 \tag{2.14}$$

which is Eq. A7.2-4 in the AISI Specification. From Eqs. (2.9), (2.13), and (2.14) it can be shown that steels with a large F_{uv}/F_{yv} ratio have more ability to strain-harden than those with a small ratio. Also, the smaller the radius-to-thickness ratio (R/t), the larger is the enhancement of the yield point. The formulae were calibrated experimentally by the choice of b and m in Eqs. (2.11) and (2.12) for

(a) $R/t < 7$
(b) $F_{uv}/F_{yv} \geq 1.2$
(c) Minimum bend angle of 60°

The formulae are not applicable outside this range.

The average yield stress of a section including the effect of cold work in the corners can be simply computed using the sum of the products of the areas of the flats and corners with their respective yield strengths. These are taken as the virgin yield point (F_{yv}) and the enhanced yield point (F_{yc}), respectively, as set out in the AISI Specification.

2.6 FRACTURE TOUGHNESS

In some instances, structures which have been designed according to proper static design sections fail by rapid fracture in spite of a significant safety factor and without violating the design loading. The cause of this type of failure is the presence of cracks, often as a result of imperfect welding or caused by fatigue, corrosion, pitting, etc. The reason for the rapid fracture is that the applied stress and crack dimensions reach critical values (related to the "fracture toughness" of the material) at which instability leads to crack extension which may even accelerate to crack propagation at the speed of an elastic wave in the propagating medium. Thus there are three factors which come into play: the applied load, the crack dimension, and the fracture toughness of the material. The discipline of "fracture mechanics" is formulating the scientific and engineering aspects of conditions that deal with cracked structures. A brief introduction to fracture mechanics can be found in Refs. 2.10 and 2.11.

It is clear that fracture toughness relates to the amount of plastic work which is needed for crack extension. However, crack extension also relates to the elastic energy released from the loaded body as the volume in the vicinity of the cracked surfaces elastically relaxes. The energy components involved therefore are positive, plastic work done by crack extension, and negative, elastic energy release. Once critical conditions are reached, the negative term dominates and the structure undergoes sudden failure by a rapid fracture.

There are two factors in cold-formed materials which reduce their ability to resist rapid fracture. The first is the reduced material ductility, described in Section 2.3, which reduces the energy absorbed in plastically deforming and fracturing at the tip of a crack, and hence its fracture toughness. The second is the increased residual stresses in cold-formed sections, produced by the forming processes, which increases the energy released as a crack propagates. Little research appears to have been performed to date on the fracture toughness of cold-formed sections other than that discussed below, probably as a result of the fact that cold-formed sections are not frequently welded. However, where connections between cold-formed sections contain welds, designers should be aware of the greater potential for rapid fracture. More research is required in this area. Investigations of cold-formed tubes with wall thickness greater than 0.8 in. have recently been performed in Japan (Refs. 2.12, 2.13).

A study of the fracture toughness of G550 sheet steels in tension has been performed and described by Rogers and Hancock (Ref. 2.14). The Mode 1 critical stress intensity factor for 0.42-mm (0.016 in.) and 0.60-mm (0.024 in.) G550 (80 ksi yield) sheet steels was determined for the cold-rolled steel in three directions of the plane of the sheet. Single notch test specimens with fatigue cracks were loaded in tension to determine the resistance of G550 sheet steels to failure by unstable fracture in the elastic deformation range. The critical stress intensity factors obtained by testing were then used in a finite element study to determine the risk of unstable fractures in the elastic deformation zone for a range of different G550 sheet steel structural models. It was determined that the perforated coupon (Ref. 2.5) and bolted connection specimens were not at risk of fracture in the elastic zone. A parametric study of perforated sections revealed that the probability of failure by unstable fracture increases, especially for specimens loaded in the transverse direction, as G550 sheet steel specimens increase in width, as cracks propagate, and for specimens loaded in a localized manner.

REFERENCES

2.1 American Society for Testing Materials, Standard Test Methods and Definition for Mechanical Testing of Steel Products, ASTM A370-95, 1995.

2.2 Dhalla, A. K. and Winter, G., Steel ductility measurements, Journal of the Structural Division, ASCE, Vol. 100, No. ST2, Feb. 1974, pp. 427–444.

2.3 Dhalla, A. K. and Winter, G., Suggested steel ductility requirements, Journal of the Structural Division, ASCE, Vol. 100, No. ST2, Feb. 1974, pp. 445–462.

2.4 Standards Australia, Steel Sheet and Strip-Hot-Dipped Zinc-Coated or Aluminium Zinc-Coated, AS 1397–1993, 1993.

2.5 Rogers, C. A. and Hancock, G. J., Ductility of G550 sheet steels in tension, Journal of Structural Engineering, ASCE, Vol. 123, No. 12, 1997, pp. 1586–1594.

2.6 Chajes, A., Britvec, S. J., and Winter, G., Effects of cold-straining on structural sheet steels, Journal of the Structural Division, ASCE, Vol. 89, No. ST2, April 1963, pp. 1–32.

2.7 Abel, A., Historical perspectives and some of the main features of the Bauschinger effect, Materials Forum, Vol. 10, No. 1, First Quarter, 1987.

2.8 Nadai, A., Theory of Flow and Fracture in Solids, Vol. 1, McGraw-Hill, New York, 1950.

2.9 Karren, K. W., Corner properties of cold-formed steel shapes, Journal of the Structural Division, ASCE, Vol. 93, No. ST1, Feb. 1967, pp. 401–432.

2.10 Chipperfield, C. G., Fracture Mechanics, Metals Australasia, March 1980, pp. 14–16.

2.11 Abel, A., To live with cracks, an introduction to concepts of fracture mechanics, Australian Welding Journal, March/April 1977, pp. 7–9.

2.12 Toyoda, M., Hagiwara, Y., Kagawa, H., and Nakano, Y., Deformability of Cold Formed Heavy Gauge RHS: Deformations and Fracture of Columns under Monotonic and Cyclic Bending Load, Tubular Structures, V, E & F.N. Spon, 1993, pp. 143–150.

2.13   Kikukawa, S., Okamoto, H., Sakae, K., Nakamura, H., and Akiyama, H., Deformability of Heavy Gauge RHS: Experimental Investigation, Tubular Structures V, E. & F.N. Spon, 1993, pp. 151–160.

2.14   Rogers, C. A. and Hancock, G. J., Fracture toughness of G550 sheet steels subjected to tension, Journal of Constructional Steel Research, Vol. 57, No. 1, 2000, pp. 71–89.

3

Buckling Modes of Thin-Walled Members in Compression and Bending

3.1 INTRODUCTION TO THE FINITE STRIP METHOD

The finite strip method of buckling analysis of thin-walled sections is a very efficient tool for investigating the buckling behavior of cold-formed members in compression and bending. The buckling modes which are calculated by the analysis can be drawn by using computer graphics, and consequently it is a useful method for demonstrating the different modes of buckling of thin-walled members. It is the purpose of this chapter to demonstrate the finite strip buckling analysis to describe generally the different modes of buckling of cold-formed members in compression and bending. Although this description is not central to the application of the design methods described in later chapters, it facilitates an understanding of these methods. In addition, the finite strip method of analysis can be used to give more accurate values of the local buckling and distortional buckling stresses than is available by simple hand methods. The AISI Specification does not explicitly allow the use of finite strip analyses to determine section strength. However, the Specification Committee is presently (2000) debating a change called the *direct strength method*, explained in Chapter 12. The Australian/New Zealand Standard (Ref. 1.4) permits the use of more accurate methods such as the finite strip method.

The *semianalytical finite strip method* used in this book is the same as that described by Cheung (Ref. 3.1) for the stress analysis of folded plate systems and subsequently developed by Przmieniecki (Ref. 3.2) for the local buckling analysis of thin-walled cross sections. Plank and Wittrick (Ref. 3.3) incorporated membrane buckling displacements in addition to plate flexural displacements to permit the study of a wide range of buckling modes ranging from local through distortional to flexural and torsional-flexural. An alternative method called the *spline finite strip method* is a development of the semianalytical finite strip method. It can be used to account for nonsimple end boundary conditions and was developed for buckling analyses of thin flat-walled structures by Lau and Hancock (Ref. 3.4). A brief comparison of the two methods when applied to a channel section of fixed length is given in Section 3.2.3.

The *semi-analytical finite strip method** involves subdividing a thin-walled member, such as the edge-stiffened plate in Figure 3.1a, into longitudinal strips. Each strip is assumed to be free to deform both in its plane (membrane displacements) and out of its plane (flexural

---

*A computer program THIN-WALL has been developed at the University of Sydney to perform a finite strip buckling analysis of thin-walled Sections under compression and bending. Computer program CU-FSM has been developed at Cornell University for finite strip buckling analyses.

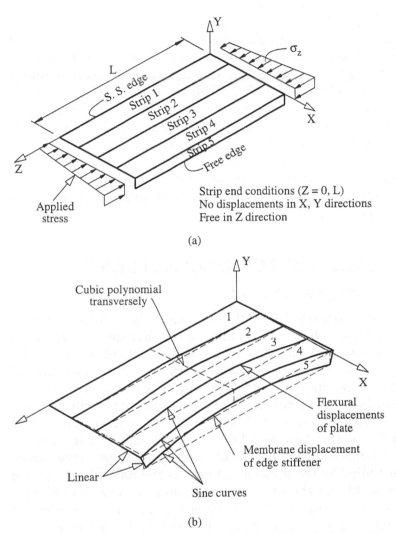

Strip end conditions (Z = 0, L)
No displacements in X, Y directions
Free in Z direction

(a)

(b)

FIGURE 3.1   Finite strip analysis of edge stiffened plate: (a) strip subdivision; (b) membrane and flexural buckling displacements.

displacements) in a single half-sine wave over the length of the section being analyzed as shown in Figure 3.1b. The ends of the section under study are free to deform long-itudinally but are prevented from deforming in a cross-

59

sectional plane. The buckling modes computed are for a single buckle half-wavelength. Details of the analytical method and its application to cases where multiple half-wavelengths occur within the length of a section under study are given in Ref. 3.5. Each strip in the cross section is assumed to be subjected to a longitudinal compressive stress (σ_z) which is uniform along the length of the strip and varies linearly from one nodal line to the other nodal line, as shown in Figure 3.1a. This allows the section under study to be subjected to a range of longitudinal stress distributions varying from pure compression to pure bending.

3.2 SINGLY SYMMETRIC COLUMN STUDY

3.2.1 Unlipped Channel

To demonstrate the different ways in which a singly symmetric channel column may buckle under both concentric and eccentric loads, the results of a semianalytical finite strip buckling analysis of an unlipped channel of depth 6 in., flange width 2 in., and thickness 0.125 in. are described and discussed. A stability analysis of the channel in Figure 3.2 subjected to a uniform compressive stress produces the two graphs in Figure 3.3. These graphs represent the buckling load (uniform compressive stress multiplied by the gross area) versus the half-wavelength of the buckle for the first two modes of buckling. A minimum (point A) occurs in the lower curve at a half-wavelength equal to approximately 6.3 in. and corresponds to local buckling in the symmetrical mode shown. Similarly at point B in the upper curve, a minimum occurs at a half-wavelength equal to approximately 4 in., which corresponds to the second mode of local buckling with the antisymmetrical mode shape shown.

At half-wavelength equal to 59 in., the first and second modes of buckling (points C and D respectively) have the

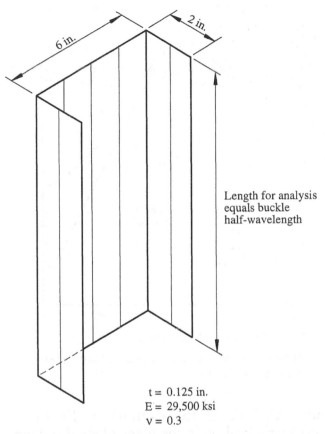

6 in.

2 in.

Length for analysis
equals buckle
half-wavelength

t = 0.125 in.
E = 29,500 ksi
v = 0.3

FIGURE 3.2   Finite strip subdivision of unlipped channel.

mode shapes shown in Figure 3.3. The lower point (C) at a
critical load of 33.5 kips (27.5 ksi) corresponds to flexural
buckling about the *y*-axis. The upper point (D) at a critical
load of 43 kips is for torsional-flexural buckling about
the *x*-axis, which is the axis of symmetry. Hence, when a
column of this section geometry is subjected to concentric
loading between simple supports spaced 59 in. apart and
which prevent rotation about the longitudinal axis, it will
buckle in the flexural mode and not in the torsional flexural
mode.

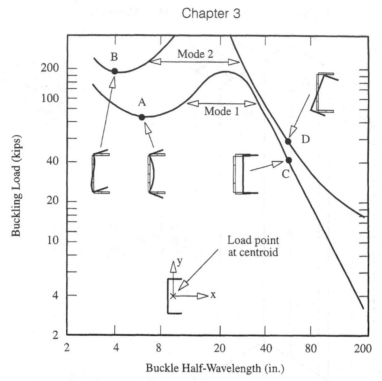

FIGURE 3.3 Unlipped channel section buckling load versus half-wavelength for concentric compression.

A stability analysis of the channel in Figure 3.2 subjected to eccentric loading in the plane of single symmetry, so that the load is located in line with the two flange tips and midway between them, produces the two graphs in Figure 3.4. These graphs represent the buckling load versus buckle half-wavelength of the section under eccentric compression for the first two modes. The curves are similar to those in Figure 3.3 except for two significant differences.

First, the torsional-flexural buckling curve, which is the upper curve in Figure 3.3, is now the lower curve in Figure 3.4 in the range 21–130 in. This means that for a column subjected to eccentric loading between simple supports spaced 59 in. apart, where the load is located

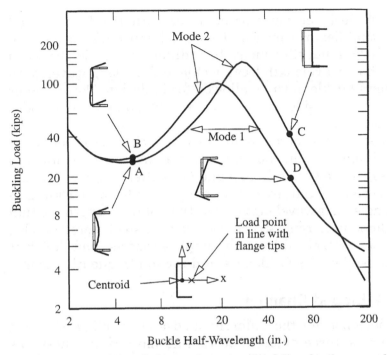

FIGURE 3.4 Unlipped channel section buckling load versus half-wavelength for eccentric compression.

midway between the flange tips, the column will buckle at a lower load in the torsional-flexural mode (point D) than in the flexural mode (point C). For half-wavelengths greater than 40 in., the flexural mode curve through point C is identically placed in Figures 3.3 and 3.4, and only the position of the torsional-flexural curve has been altered.

Second, the local buckling load (point A) has been lowered from 63 kips (51.6 ksi) (average stress) for the concentric compression in Figure 3.3 to 21.5 kips (17.7 ksi average stress) for the eccentric compression in Figure 3.4. The first two modes of local buckling (points A and B) occur at nearly identical loads for the case of eccentric compression in Figure 3.4.

A designer must account for both the flexural and torsional-flexural modes of long column buckling as well as the local buckling mode. In addition, interaction between the short-wavelength local buckle and the long-wavelength column buckle must be allowed for in design. The designer must also consider the effect of yielding on all three modes of buckling.

In the case of torsional-flexural buckling, the boundary conditions used in the finite strip analysis are identical to those used by Timoshenko and Gere (Ref. 3.6) for a simply supported column free to warp at its ends and buckling in a single half-wavelength over the column length. Actual columns in normal structures, such as steel storage racks, are not restrained by such simple supports, and different effective lengths for flexure and torsion frequently occur.

3.2.2 Lipped Channel

To demonstrate the different modes of buckling of lipped channels, three different section geometries are chosen as shown in Figure 3.5. The three sections chosen are of the type used in steel storage rack columns where additional flanges (called rear flanges) may be added to allow bolting of bracing to the channels as shown in Figure 1.4a. The basic profile (called Section 1) is the lipped channel with

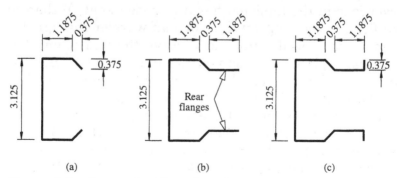

FIGURE 3.5   Geometry of lipped sections: (a) section 1 ($t = 0.060$ in.); (b) section 2 ($t = 0.060$ in.); (c) section 3 ($t = 0.060$ in.).

sloping lip stiffeners shown in Figure 3.5a. Sloping lip stiffeners were chosen so that Section 2 (shown in Figure 3.5b) is identical to Section 1 except for the inclusion of the rear flanges. The geometry of Section 2 was chosen to be representative of a typical channel with rear flanges. Section 3 (shown in Figure 3.5c) is identical to Section 2 except for the inclusion of additional lip stiffeners on the rear flanges. All three sections studied are of the same overall depth equal to 3.125 in. and plate thickness equal to 0.060 in. The sections were analyzed with a semianalytical finite strip buckling analysis.

The results of a stability analysis of Section 1, subjected to uniform compression, are shown in Figure 3.6 as the buckling stress versus buckle half-wavelength. A minimum (point A) occurs in the curve at a half-wavelength of 2.6 in. and represents *local* buckling in the mode shown. The local mode consists mainly of deformation of the web element without movement of the line junction

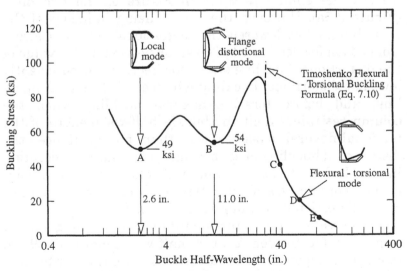

FIGURE 3.6 Section 1—Buckling stress versus half-wavelength for concentric compression.

65

between the flange and lip stiffener. A minimum also occurs at a point B at a half-wavelength of 11.0 in. in the mode shown. This mode is a *flange distortional* buckling mode since movement of the line junction between the flange and lip stiffener occurs without a rigid body rotation or translation of the cross section. In some papers, this mode is called a *local-torsional* mode. The distortional buckling stress at point B is slightly higher than the local buckling stress at point A so that when a long-length fully braced section is subjected to compression, it is likely to undergo local buckling in preference to distortional buckling. As for the unlipped channel described previously, the section buckles in a flexural or torsional-flexural buckling mode at long wavelengths, such as at points C, D, and E in Figure 3.6. For this particular section, torsional-flexural buckling occurs at half-wavelengths up to approximately 71 in., beyond which flexural buckling occurs.

The results of a stability analysis of Section 2 (Figure 3.5b) subjected to uniform compression are shown in Figure 3.7 and are similar to those in Figure 3.6 for Section 1 (Figure 3.5a). However, the destabilizing influence of the rear flanges has lowered the distortional buckling stress from 54 ksi for Section 1 to 26 ksi for Section 2. The value of the distortional buckling stress for Section 2 is sufficiently low that distortional buckling will occur before local buckling, and before torsional-flexural buckling when the column restraints limit the buckle half-wavelength of the torsional-flexural mode to less than 55 in. Hence, the distortional buckling mode becomes a serious consideration in the design of such a column. A photograph of a cold-formed column of a similar type to Section 2 and undergoing distortional buckling is shown in Figure 3.8.

The value of the torsional-flexural buckling stress, computed for Section 2 at a half-wavelength of 59 in. (point D in Figure 3.7), is 13% higher than that for Section 1 at the same half-wavelength (point D in Figure 3.6) since the section with rear flanges is more efficient at resisting

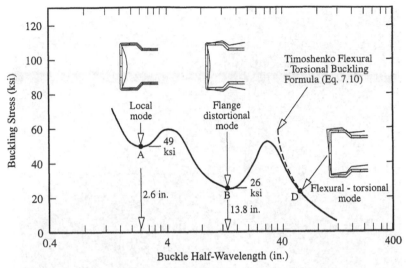

FIGURE 3.7   Section 2—Buckling stress versus half-wavelength for concentric compression.

torsional-flexural buckling as a result of a fourfold increase in the section warping constant.

The results of a stability analysis of Section 3 (Figure 3.5c) to uniform compression are shown in Figure 3.9. This graph is also similar in form to that drawn in Figures 3.6 and 3.7. However, the destabilizing influence of the rear flanges in the distortional mode has been significantly decreased by the inclusion of the additional stiffening lips on the rear flanges. The distortional buckling stress at point B has been increased from 26 ksi for Section 2 to 39 ksi for Section 3. Hence, the distortional buckling mode will be a much less serious consideration in the design of a column with additional lip stiffeners than it would be for the section with unstiffened rear flanges.

Unfortunately, inclusion of the additional lip stiffeners in Section 3 results in a reduction of the torsional-flexural buckling stress by 15% below that for Section 2, mainly as a consequence of the increased distance in Section 3 between the shear center and the centroid. Hence, the inclusion of

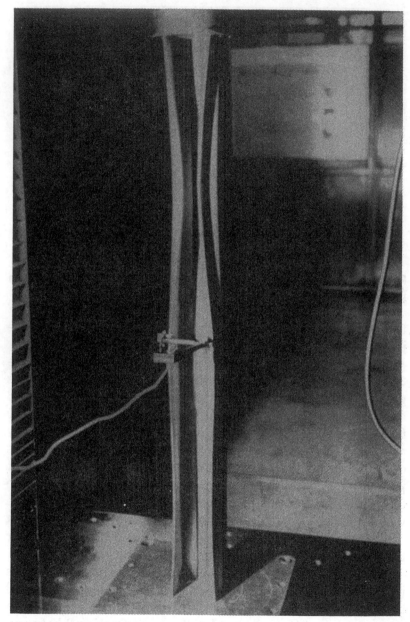

FIGURE 3.8 Flange distortional buckle of lipped channel column—Section 2.

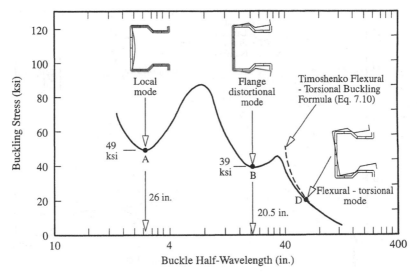

FIGURE 3.9   Section 3—Buckling stress versus half-wavelength for concentric compression.

the additional lip stiffeners will be detrimental to the strength of the section where torsional-flexural buckling controls the design. Additional information on the distortional mode of buckling, including design formulas, is given in Refs. 3.7, 3.8, and 3.9.

3.2.3 Lipped Channel (Fixed Ended)

The spline finite strip method of buckling analysis (Ref. 3.4) uses spline functions in the longitudinal direction in place of the single half-sine wave over the length of the section as demonstrated in Figure 3.1b. The advantage of the spline finite strip analysis is that boundary conditions other than simple supports can be investigated. However, the spline finite strip analysis uses a considerably greater number of degrees of freedom and takes much longer to solve a given problem than the semianalytical finite strip method.

To investigate the effect of fixed-ended boundary conditions on the distortional buckling of a simple lipped

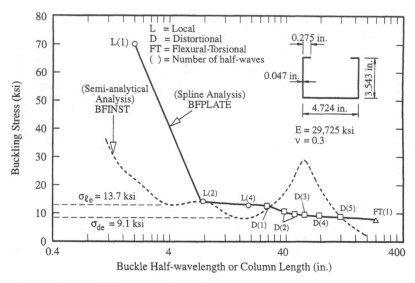

FIGURE 3.10   Finite strip buckling analyses.

channel, both types of finite strip analysis were performed for the channel section in Figure 3.10. The results of buckling analyses of the lipped channel section are shown in Figure 3.10, where BFINST denotes the semianalytical finite strip analysis and BFPLATE denotes the spline finite strip analysis. The BFINST analysis gives the elastic local (σ_{le}) and elastic distortional (σ_{de}) buckling stresses at given half-wavelengths, whereas the BFPLATE analysis gives the actual buckling stress of a given length of section between ends that may be fixed. The abscissa in Figure 3.10 for the BFINST analysis indicates one half-wave whereas for the BFPLATE analysis the abscissa indicates the real column length of the section. The different marks indicate local (L), distortional (D), and overall buckling (FT), respectively. It can be observed in Figure 3.10 that where the sections buckle in multiple distortional half-waves the change in the buckling stress in the distortional mode with increasing section length reduces as the effect of the end conditions reduces, and eventually approaches the value produced by the semianalytical finite strip method.

3.3 PURLIN SECTION STUDY

3.3.1 Channel Section

To demonstrate the different ways in which a channel purlin may buckle when subjected to major axis bending moment as shown in Figure 3.11, a finite strip buckling analysis of a channel purlin of depth 6 in., flange width 2.5 in., lip size 0.55 in., and plate thickness 0.060 in. was performed and the results are shown in Figure 3.12. This graph is similar to that for the lip-stiffened channel in Figure 3.6 except for the shape of the buckling modes and their nomenclature.

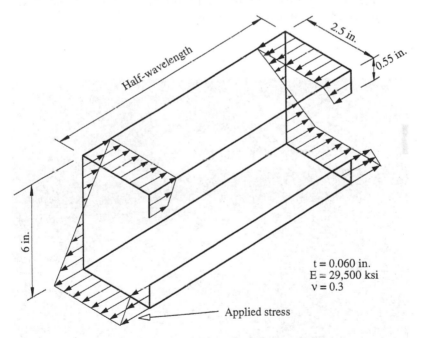

FIGURE 3.11 Channel section purlin subjected to major axis bending moment.

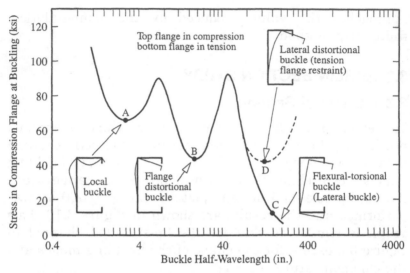

FIGURE 3.12   Channel section purlin buckling stress versus half-wavelength for major axis bending.

FIGURE 3.13   Flange distortional buckle of Z-section purlin.

The first minimum point (A) is again a *local* buckling mode, which now involves only the web, the compression flange, and its lip stiffener. The second minimum point (B) is associated with a mode of buckling where the compression flange and lip rotate about the flange web junction with some elastic restraint to rotation provided by the web. This mode of buckling is called a *flange distortional* buckling mode and is shown in a test specimen of a Z-section in Figure 3.13. The value of the buckling stress is highly dependent upon the size of the lip stiffener.

At long wavelengths (point C) where the purlin is unrestrained, a torsional-flexural buckle occurs which is often called a *lateral* buckle. However, if the tension flange is subjected to a torsional restraint such as may be provided by sheeting screw fastened to the tension flange, then a *lateral distortional* buckle will occur at a minimum half-wavelength of approximately 160 in., as shown at point D in Figure 3.12. The value of the minimum buckling stress and its half-wavelength depend upon the degree of torsional restraint provided to the tension flange.

3.3.2 Z-Section

A similar study to that of the channel section in bending was performed for the two Z-sections in Figure 3.14. The first section contains a lip stiffener perpendicular to the flange, and the second has a lip stiffener located at an angle of 45° to the flange. In Figure 3.14, the buckling stresses have been computed for buckle half-wavelengths up to 40 in., so only the local and distortional buckling modes have been investigated.

As for the channel section study described above, the distortional buckling stresses are significantly lower than the local buckling stresses for both Z-sections. For the lip stiffener turned at 45° to the flange, the distortional buckling stress is reduced by 19% compared with the value for the lip

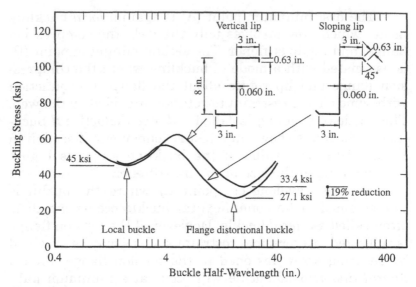

FIGURE 3.14   Z-section purlin—buckling stress versus half-wavelength for bending about a horizontal axis.

stiffener perpendicular to the flange, thus indicating a potential failure mode of purlins with sloping lip stiffeners.

3.4 HOLLOW FLANGE BEAM IN BENDING

The hollow flange beam section in Figure 3.15 was investigated by a semianalytical finite strip buckling analysis. Two sections have been analyzed to demonstrate the effect of the electric resistance weld (ERW) on the section buckling behavior. These are the section with closed flanges (HBS1) and the section with open flanges (HBS2). Figure 3.15 shows graphs of buckling stress versus buckle half-wavelengths for the two sections subjected to pure bending about their major principal axes so that their top flanges are in compression and their bottom flanges are in tension as in a conventional beam. The buckling stress is the value of the stress in the compression flange farthest away from the bending axis when the section undergoes elastic buckling.

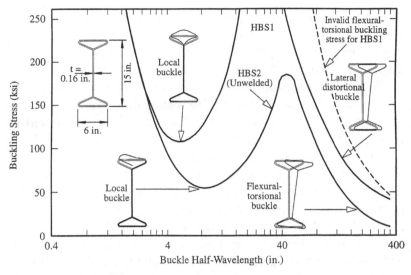

FIGURE 3.15 Hollow flange beam—buckling stress versus half-wavelength for major axis bending.

At short half-wavelengths (2–10 in. in Figure 3.15), the effect of welding the flange to the abutting web clearly demonstrates the changed buckling mode from local buckling in the unattached flange for HBS2 to local buckling of the top flange at a higher stress for HBS1.

At long half-wavelengths (80–400 in. in Figure 3.15), the increased torsional rigidity of the flanges has increased the buckling stress, with the increase exceeding 100% for lengths greater than 200 in. The mode of buckling at a half-wavelength of 200 in. for the open section (HBS2) is a conventional torsional-flexural buckle, of the type described in Ref. 3.6. The torsional-flexural buckling mode of the open HBS2 section involves longitudinal displacements of the cross-section (called warping displacements) such that the longitudinal displacements at the free edges of the strips are different from the longitudinal displacements of the web at the points where the free edges abut the web.

The mode of buckling at 200 in. for the section with closed flanges (HBS1) shows a new type of buckling mode not previously described for sections of this type. It involves lateral bending of the two flanges, one more than the other, with the flanges substantially untwisted as a result of their increased torsional rigidity. The web is distorted as a result of the relative movement of the flanges. The mode is called *lateral distortional buckling* and was first reported for sections of this type in Ref. 3.10. It has a substantially increased buckling stress value over that of the torsional-flexural buckling of the open section HBS2. It is not valid to compute the torsional-flexural buckling capacity for HBS1 using Ref. 3.6 because this would produce erroneous results, as shown by the dashed line in Figure 3.15. The differential warping displacements where the free edges of the flange abut the web in the open HBS2 section are eliminated by the welds in the closed HBS1 section. Consequently, the enhanced lateral distortional buckling stress of the HBS1 section is obtained entirely as a result of the welds. An analytical formula for computing lateral distortional buckling is given in Ref. 3.11.

REFERENCES

3.1 Cheung, Y. K., Finite Strip Method in Structural Analysis, Pergamon Press, New York, 1976.

3.2 Przmieniecki, J. S. D., Finite element structural analysis of local instability, Journal of the American Institute of Aeronautics and Astronautics, Vol. 11, No. 1, 1973.

3.3 Plank, R. J. and Wittrick, W. H., Buckling under combined loading of thin, flat-walled structures by a complex finite strip method, International Journal For Numerical Methods in Engineering, Vol. 8, No. 2, 1974, pp. 323–329.

3.4 Lau, S. C. R. and Hancock, G. J., Buckling of thin flat-walled structures by a spline finite strip method,

Thin-Walled Structures, 1986, Vol. 4, No. 4, pp. 269–294.

3.5 Hancock, G. J., Local, distortional and lateral buckling of I-beams, Journal of the Structural Division, ASCE, Vol. 104, No. ST11, Nov. 1978, pp. 1787–1798.

3.6 Timoshenko, S. P. and Gere, J. M., Theory of Elastic Stability, McGraw-Hill, New York, 1959.

3.7 Hancock, G. J., Distortional buckling of steel storage rack columns, Journal of Structural Engineering, ASCE, Vol. 111, No. 12, Dec. 1985.

3.8 Lau, S. C. W. and Hancock, G. J., Distortional buckling formulas for channel columns, Journal of Structural Engineering, ASCE, Vol. 113, No. 5, May 1987.

3.9 Lau, S. C. W. and Hancock, G. J. Distortional Buckling Tests of Cold-Formed Channel Sections, Proceedings, Ninth International Specialty Conference on Cold-Formed Steel Structures, St. Louis, Missouri, Nov. 1988.

3.10 Centre for Advanced Structural Engineering, Distortional Buckling of Hollow Flange Beam Sections, Investigation Report S704, University of Sydney, Feb. 1989.

3.11 Pi, Y. K. and Trahair, N. S., Lateral-distortional buckling of hollow flange beams, Journal of Structural Engineering, ASCE, Vol. 123, No. 6, 1997, pp. 695–702.

4

Stiffened and Unstiffened Compression Elements

4.1 LOCAL BUCKLING

Local buckling involves flexural displacements of the plate elements, with the line junctions between plate elements remaining straight, as shown in Figures 3.6, 3.7, 3.9, and 3.12. The elastic critical stress for local buckling has been extensively investigated and summarized by Timoshenko and Gere (Ref. 3.6), Bleich (Ref. 4.1), Bulson (Ref. 4.2), and Allen and Bulson (Ref. 4.3). The elastic critical stress for local buckling of a plate element in compression, bending, or shear is given by

$$f_{ol} = \frac{k\pi^2 E}{12(1 - v^2)} \left(\frac{t}{w}\right)^2 \tag{4.1}$$

where k is called the plate local buckling coefficient and depends upon the support conditions, and w/t is the plate slenderness, which is the plate width (w) divided by the plate thickness (t).

A summary of plate local buckling coefficients with the corresponding half-wavelengths of the local buckles is shown in Figure 4.1. For example, a plate with simply supported edges on all four sides and subjected to uniform compression will buckle at a half-wavelength equal to w with $k = 4.0$. A plate with one longitudinal edge free and the other simply supported will buckle at a half-wavelength equal to the plate length (L), and if this is sufficiently long then $k = 0.425$. However, if the half-wavelength of the buckle is restricted to a length equal to twice its width ($L = 2w$), then $k \approx 0.675$, as set out in Figure 4.1.

For the unlipped channel in Figure 3.2 and subjected to uniform compression, if each flange and the web are analyzed in isolation by ignoring the rotational restraints provided by the adjacent elements, then the buckling coefficients are $k = 0.43$ for the flanges and $k = 4.0$ for the web. These produce buckling stresses of 48.7 ksi for the flanges at an infinite half-wavelength and 48.4 ksi for the web at a half-wavelength of 5.866 in. A finite strip buckling analysis shows that the three elements buckle simultaneously at the same half-wavelength of approximately 6.3 in. at a compressive stress of 50.8 ksi. This stress is higher than either of the stresses for the isolated elements because of the changes required to make the half-wavelengths compatible.

For the lipped channel purlin in Figure 3.11, the buckling coefficients for the web in bending, the flange in uniform compression, and the lip in near uniform compressin are 23.9, 4.0, and 0.43, respectively. The corresponding buckling stresses are 63.8 ksi, 58.6 ksi, and 142.9 ksi, respectively. In this case, a finite strip buckling analysis shows that the three elements buckle at a stress and half-wavelength of 65.3 ksi and 3.54 in., respectively.

For both of the cases described above, a designer would not normally have access to an interaction buckling analysis and would use the lowest value of buckling stress in the cross section, considering the individual elements in isolation.

Case	Boundary Conditions	Loading	Buckling Coefficient (k)	Half - Wavelength
1	S.S / S.S / S.S / S.S	Uniform Compression	4.0	w
2	S.S / Built-in / S.S / Built-in	Uniform Compression	6.97	0.66w
3	S.S / S.S / S.S / Free	Uniform Compression	0.425 0.675	$L = \infty$ $L = 2w$
4	S.S / Built-in / S.S / Free	Uniform Compression	1.247	1.636w
5	S.S / S.S / S.S / S.S	Pure Bending	23.9	0.7w
6	S.S / S.S / S.S / S.S	Bending + Compression	7.81	w
7	S.S / Free / S.S / S.S	Bending + Compression	0.57	$L = \infty$
8	S.S / S.S / S.S / S.S	Pure Shear	5.35 9.35	$L = \infty$ $L = w$

L = Plate length, w = Plate width

FIGURE 4.1   Plate buckling coefficients.

4.2 POSTBUCKLING OF PLATE ELEMENTS IN COMPRESSION

Local buckling does not normally result in failure of the section as does flexural (Euler) buckling in a column. A plate subjected to uniform compressive strain between rigid frictionless platens will deform after buckling, as shown in

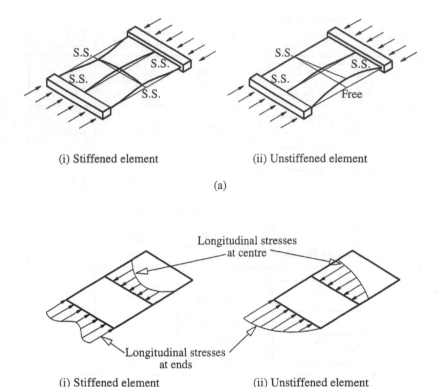

(i) Stiffened element   (ii) Unstiffened element

(a)

(i) Stiffened element   (ii) Unstiffened element

(b)

FIGURE 4.2   Postbuckled plates: (a) deformations; (b) stresses.

Figure 4.2a, and will redistribute the longitudinal membrane stresses from uniform compression to those shown in Figure 4.2b. This will occur irrespective of whether the plate is a stiffened or an unstiffened element. The plate element will continue to carry load, although with a stiffness reduced to 40.8% of the initial linear elastic value for a square stiffened element and to 44.4% for a square unstiffened element (Ref. 4.2). However, the line of action of the compressive force in an unstiffened element will move toward the stiffened edge in the postbuckling range.

The theoretical analysis of postbuckling and failure of plates is extremely difficult and generally requires a computer analysis to achieve an accurate solution. To avoid this complex analysis in design, von Karman (Ref. 4.4) suggested that the stress distribution at the central section of a stiffened plate be replaced by two widths ($b/2$) on each side of the plate, each subjected to a uniform stress (f_{max}) as shown in Figure 4.3a, such that $f_{max}bt$ equals the actual load in the plate. Von Karman called b the effective width.

Von Karman also suggested that the two strips be considered as a rectangular plate of width b and that when the elastic critical stress of this plate is equal to the yield strength (F_y) of the material, then failure of the plate occurs. From Eq. (4.1),

$$F_y = \frac{k\pi^2 E}{12(1 - v^2)} \left(\frac{t}{b}\right)^2 \tag{4.2}$$

Dividing Eq. (4.1) by Eq. (4.2) produces

$$\frac{b}{w} = \sqrt{\frac{f_{ol}}{F_y}} \tag{4.3}$$

Equation (4.3) is von Karman's formula for effective width and can be used for design. Although von Karman only suggested the effective width formula for stiffened elements, it appears to work satisfactorily for the effective width of the unstiffened element in Figure 4.3b. In this

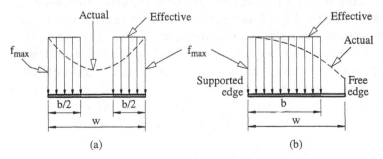

FIGURE 4.3   Effective stress distributions: (a) stiffened element; (b) unstiffened element.

case, the full effective width (b) is located adjacent to the supported edge.

4.3 EFFECTIVE WIDTH FORMULAE FOR IMPERFECT ELEMENTS IN PURE COMPRESSION

Compression elements in cold-formed sections contain geometric imperfections as well as residual stresses from the cold-forming operation. Consequently, the von Karman formula for effective width needs to be modified to account for the reduction in strength resulting from these imperfections. Winter has proposed and verified experimentally (Refs. 4.5, 4.6) the following effective width formulae for stiffened [Eq. (4.4)] and unstiffened [Eq. (4.5)] compression elements:

$$\frac{b}{w} = \sqrt{\frac{f_{ol}}{F_y}}\left(1 - 0.22\sqrt{\frac{f_{ol}}{F_y}}\right) \tag{4.4}$$

$$\frac{b}{w} = 1.19\sqrt{\frac{f_{ol}}{F_y}}\left(1 - 0.298\sqrt{\frac{f_{ol}}{F_y}}\right) \tag{4.5}$$

A comparison of test results of cold-formed sections which had unstiffened elements (Ref. 4.7) with Eqs. (4.4) and (4.5) is given in Figure 4.4. For slender plates ($\sqrt{F_y/f_{ol}} > 2.0$), Eq. (4.5) for unstiffened elements gives the better approximation to the test results. However, for stockier plates, Eq. (4.4) provides a more accurate estimate of the plate strength. Consequently, in the AISI Specification, Eq. (4.4) has been used for both stiffened and unstiffened compression elements. This also has the advantage of simplicity since one equation can be used for both types of elements. The resulting equation for effective width produced by substituting Eq. (4.1) in Eq. (4.4) is

$$\frac{b}{t} = 428\sqrt{\frac{k}{F_y}}\left(1 - \frac{93.5}{w/t}\sqrt{\frac{k}{F_y}}\right) \tag{4.6}$$

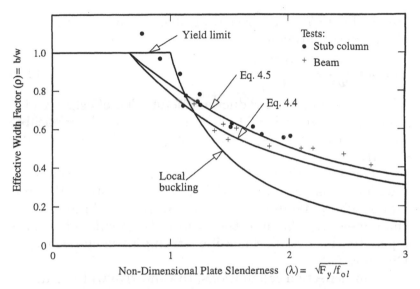

FIGURE 4.4   Tests of unstiffened compressioin elements.

It has been proven that Eq. (4.6) applies at stresses less than yield, so F_y can be replaced by f to produce

$$\frac{b}{t} = 428\sqrt{\frac{k}{f}}\left(1 - \frac{93.5}{w/t}\sqrt{\frac{k}{f}}\right) \qquad (4.7)$$

where f is the design stress in the compression element calculated on the basis of the effective design width.

For stiffened compression elements, k is 4.0, and for unstiffened elements, k is 0.43.

In the AISI Specification, λ (slenderness factor) is used to represent nondimensional plate slenderness at the stress f so that

$$\lambda = \sqrt{\frac{f}{f_{ol}}} = \frac{1.052}{\sqrt{k}}\left(\frac{w}{t}\right)\sqrt{\frac{f}{E}} \qquad (4.8)$$

85

Hence, Eq. (4.7) can be rearranged as given in Section B2.1 of the Specification as

$$\rho = \frac{b}{w} = \frac{1 - 0.22/\lambda}{\lambda} \tag{4.9}$$

where ρ is called the reduction factor. The effective width (b) equals w when $\lambda = 0.673$, so Eq. (4.9) is only applicable for $\lambda > 0.673$. For $\lambda \leq 0.673$, the plate element is fully effective.

For strength limit states, where failure occurs by yielding, b is calculated with $f = F_y$. For strength limit states where failure occurs by overall buckling rather than yielding, the effective width is calculated with f equal to the overall buckling strength, as described in Chapters 5–8.

For deflection calculations, the effective width is determined by using the effective width formula with $f = f_d$, the stress in the compression flanges at the load for which deflections are determined. The resulting equation is given in Section B2.1 of the AISI Specification.

Two procedures are given in Section B2.1. Procedure I simply substitutes f_d for f in Eq. (4.9) and produces low estimates of effective widths and, hence, greater deflection. Procedure II is based on a study by Weng and Peköz (Ref. 4.8) which proposed more accurate effective width equations for calculating deflection. These are

$$\rho = \frac{b_{ed}}{w} = \frac{1.358 - 0.461/\lambda}{\lambda} \qquad \text{(for } 0.673 < \lambda < \lambda_c) \tag{4.10}$$

$$\rho = \frac{b_{ed}}{w}$$

$$= \frac{0.41 + 0.59\sqrt{F_y/f_d} - 0.22/\lambda}{\lambda} \qquad \text{(for } \lambda \geq \lambda_c) \tag{4.11}$$

where

$$\lambda_c = 0.256 + 0.328\left(\frac{w}{t}\right)\sqrt{\frac{F_y}{E}} \tag{4.12}$$

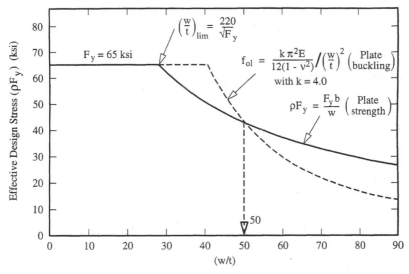

FIGURE 4.5 Effective design stress (ρF_y) on a stiffened compression element ($F_y = 65$ ksi).

When calculating λ, f_d is substituted for f, and ρ shall not exceed 1.0 for all cases.

The design curves for the effective design stress (ρF_y) in a compression element are based on the ultimate plate strength ($F_y bt$), where b is given by Eq. (4.7). The curves are drawn in Figure 4.5 for a stiffened compression element with $k = 4.0$, and in Figure 4.6 for an unstiffened compression element with $k = 0.43$. A value $f = F_y = 65$ ksi has been used in Figures 4.5 and 4.6. Although effective widths rather than effective stresses are used in the AISI Specification, these figures are useful since they give the average stress acting on the full plate element at failure. It is interesting to observe that the plate slenderness values beyond which the local buckling stress is lower than the plate strength are approximately 50 and 15 for the stiffened and unstiffened compression elements, respectively.

For uniformly compressed stiffened elements with circular holes, Eq. (4.9) can be modified as specified in

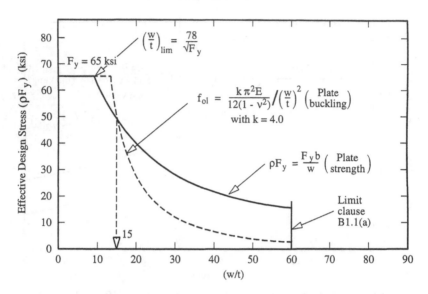

FIGURE 4.6 Effective design stress on an unstiffened (ρF_y) compression element ($F_y = 65$ ksi).

Section B2.2 of the AISI Specification. For nonslender elements where $\lambda \leq 0.673$, the hole diameter (d_h) is simply removed from the flat width (w). However, for slender elements with $\lambda > 0.673$, b is given by Eq. (4.13), such that it must be less than or equal to $w - d_h$:

$$b = w\frac{1 - 0.22/\lambda - 0.8d_h/w}{\lambda} \tag{4.13}$$

These equations are based on research by Ortiz-Colberg and Peköz (Ref. 4.9) and only apply when $w/t \leq 70$ and the center-to-center spacing of holes is greater than $0.5w$ and $3d_h$. These equations do not apply for other hole shapes such as square holes where the plates adjacent to the longitudinal edges may be regarded as unstiffened elements and probably should be designed as such.

4.4 EFFECTIVE WIDTH FORMULAE FOR IMPERFECT ELEMENTS UNDER STRESS GRADIENT

4.4.1 Stiffened Elements

In the AISI Specification, elements of this type are designed using the effective width approach in Figure 4.7a. As for a stiffened element in uniform compression, the effective width is split into two parts and is taken at the edges of the compression region, as shown in Figure 4.7a. However, the widths b_1 and b_2 are not equal as for uniformly compressed stiffened elements. Equations for the modified effective widths are given in Section B2.3 of the AISI Specification. The effective widths depend upon the stress gradient given by $\psi = f_2/f_1$. Further, the buckling coefficient (k) varies between 4.0, for pure compression, and 23.9, for pure bending as given in Figure 4.1. The value of k to be

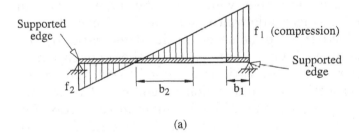

(a)

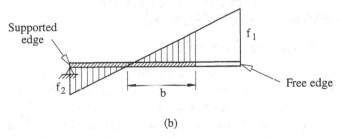

(b)

FIGURE 4.7 Effective widths of plate elements under stress gradient: (a) stiffened; (b) unstiffened.

used depends upon the stress gradient given by ψ and defined by Eq. (B2.3-4) in the AISI Specification. Further discussion and application is given in Section 6.3 of this book. The calibration of this process has been provided by Peköz (Ref. 4.10).

4.4.2 Unstiffened Elements

Unstiffened elements under stress gradient, such as flanges of channel sections bent about the minor principal axis, can be simply designed, assuming they are under uniform compression with the design stress (f) equal to the maximum stress in the element. This approach is the basic method given in Section B3.2 of the AISI Specification and is the method specified in the AISI Specification (Ref. 1.16). It assumes a buckling coefficient (k) for uniform compression of 0.43. However, a higher tier approach is given in Appendix F of AS/NZS 4600 (Ref. 1.4). This approach uses effective widths of the type shown in Figure 4.7b, where allowance is made for the portion in tension. Further, k can be based on the stress ratio (ψ) and will usually be higher than 0.43. In the application of Appendix F, the stress ratio is assumed to be that in the full (gross) section, so no iteration is required. This simplifies the process somewhat. The formulae in Appendix F of AS/NZS 4600 are based on Eurocode 3, Part 1.3 (Ref. 1.7).

4.5 EFFECTIVE WIDTH FORMULAE FOR ELEMENTS WITH STIFFENERS

4.5.1 Edge-Stiffened Elements

An edge stiffener is located along the free edge of an unstiffened plate to transform it into a stiffened element as shown in Figure 1.12b. Consequently, the local buckling coefficient is increased from 0.43 to as high as 4.0 with a corresponding increase in the plate strength. The effective width of an adequately stiffened element is as shown in

Figure 1.12c. However, a partially stiffened element may have the effective areas distributed as shown in Figure 4.8a, where the values of C_1 and C_2 depend upon the adequacy of the edge stiffener. Further, the edge stiffener itself may not be fully effective as shown by its effective width (d_s) in Figure 4.8a.

The formulae for the effective width of edge-stiffened elements are based upon Eq. (4.9). However, the adequacy of the stiffener governs the buckling coefficient to be used in Eq. (4.8). Considerable research into edge-stiffened elements has been performed by Desmond, Peköz, and Winter (Ref. 4.11) to determine the necessary equations for design. These equations are given in Section B4.2 of the AISI Specification. Three cases have been identified, as shown in Figure 4.9. Case I applies to sections which are fully effective without stiffeners, Case II applies to sections which are fully effective if the edge stiffener is adequate,

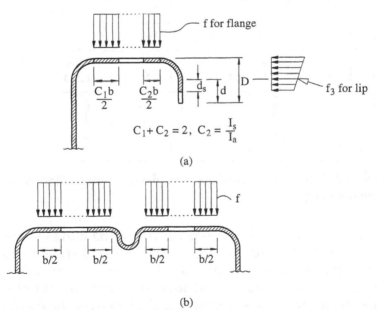

(a)

(b)

FIGURE 4.8 Effective widths for elements with stiffeners: (a) edge stiffeners; (b) intermediate stiffener.

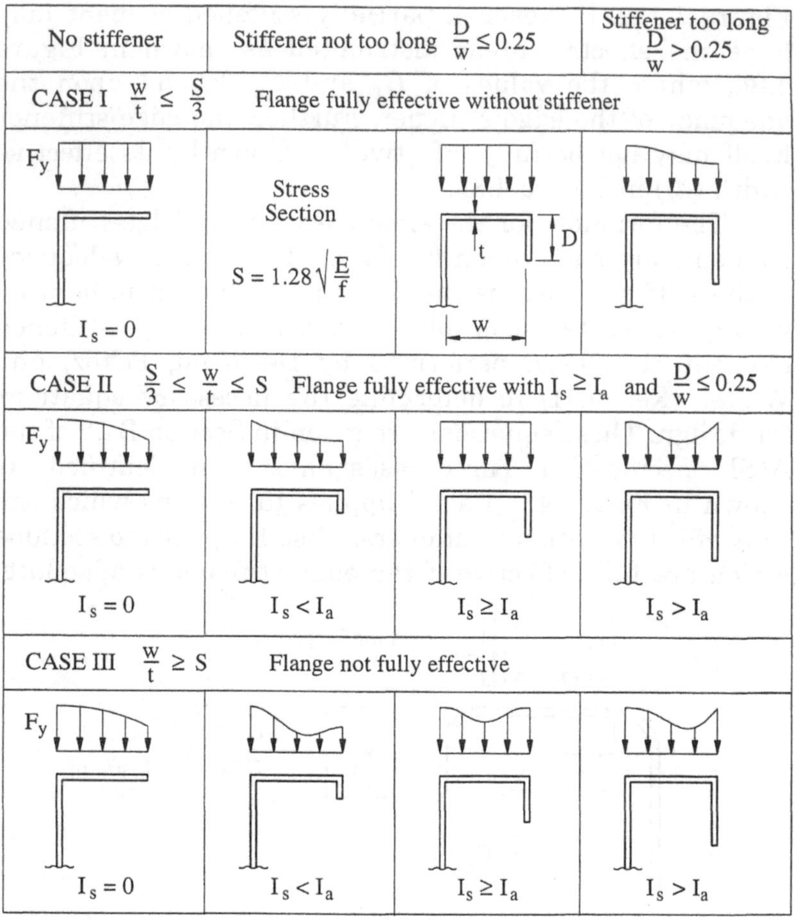

No stiffener	Stiffener not too long $\frac{D}{w} \leq 0.25$	Stiffener too long $\frac{D}{w} > 0.25$
CASE I $\frac{w}{t} \leq \frac{S}{3}$	Flange fully effective without stiffener	

F_y

Stress
Section

$S = 1.28\sqrt{\frac{E}{f}}$

$I_s = 0$

D t w

CASE II $\frac{S}{3} \leq \frac{w}{t} \leq S$	Flange fully effective with $I_s \geq I_a$ and $\frac{D}{w} \leq 0.25$		

F_y

$I_s = 0$ $I_s < I_a$ $I_s \geq I_a$ $I_s > I_a$

CASE III $\frac{w}{t} \geq S$	Flange not fully effective		

F_y

$I_s = 0$ $I_s < I_a$ $I_s \geq I_a$ $I_s > I_a$

FIGURE 4.9 Stress distributions and design criteria for edge-stiffened elements.

and Case III applies to sections for which the flange is not fully effective. Figure 4.9 shows the stress distributions which occur in the different cases. It can be observed that the stress distributions depend not only on the case but also on the adequacy of the stiffener and the length of the lip (D) relative to the flange width (w). The adequacy of the stiffener is defined by the ratio of the stiffener second moment of area (I_s) to an adequate value (I_a) specified in Section B4.2. In the AISI Specification, the stiffener second moment of area is for the flat portion (d) of the stiffener, as shown in Figure 4.8a.

In Section B4.2, if the stiffener is inadequate ($I_s/I_a < 1.0$), the effective area of the stiffener (A'_s) is further reduced to A_s, as given by Eq. (4.14) when calculating the overall effective section properties. The procedures allows for a gradual transition from stiffened to unstiffened elements without a sudden reduction in capacity if the stiffener is not adequate:

$$A_s = \left(\frac{I_s}{I_a}\right)A'_s \qquad (4.14)$$

4.5.2 Intermediate Stiffened Elements with One Intermediate Stiffener

Intermediate stiffened elements with one intermediate stiffener, as shown in Figure 4.8b, are treated in a similar manner in the AISI Specification to edge-stiffened elements using effective widths which depend upon the adequacy of the stiffener. However, the effective portions of the elements adjacent to the stiffener are distributed evenly on each side of each element as shown in Figure 4.8b. As for the edge-stiffened element, the buckling coefficient depends upon the adequacy of the stiffener as defined by I_s/I_a. Also A'_s is further reduced to A_s, as given by Eq. (4.14). The detailed design rules are given in Section B4.1 of the AISI Specification. The design rules are based on the research of Desmond, Peköz, and Winter (Ref. 4.12).

4.5.3 Intermediate Stiffeners for Edge-Stiffened Elements with an Intermediate Stiffener, and Stiffened Elements with More than One Intermediate Stiffener

Desmond, Peköz, and Winter (Refs. 4.11 and 4.12) did not include these cases, and hence design cannot be performed in the same way as described in Sections 4.5.1 and 4.5.2. The approach used in the AISI Specification is the same as previously used in the 1980 edition of the AISI Specification, where the effect of the stiffener is disregarded if the intermediate stiffener has a second moment of area less than a specified minimum.

The AISI Specification includes a method for the design of edge-stiffened elements with one or more intermediate stiffeners or stiffened elements with more than one intermediate stiffener in Section B5. Since any intermediate stiffener can be regarded as supporting the plate on both its sides, the minimum value of the second moment of area of the stiffener ($I_{\min}$) needs to be sufficient to prevent deformation of the edges of the adjacent plate. Equation (4.15) gives this minimum value:

$$I_{\min} = 3.66t^4 \sqrt{\left(\frac{w}{t}\right)^2 - \frac{0.136E}{F_y}} \geq 18.4t^4 \tag{4.15}$$

In addition, Section B5(a) specifies that if the spacing of intermediate stiffeners between two webs is such that the elements between stiffeners are wider than the limiting slenderness specified in Section B2.1 so that they are not fully effective, then only the two intermediate stiffeners (those nearest each web) shall be considered effective. This is a consequence of a reduction in shear transfer in wide unstiffened elements which are locally buckled. This effect is similar to the shear lag effect normally occurring in wide elements except that it is a result of local buckling rather than simple shear straining deformation.

For the purpose of design of a multiple stiffened element where all flat elements are fully effective, the entire segment is replaced by an equivalent single flat element whose width is equal to the overall width between

94

edge stiffeners (b_o), as shown in Figure 1.12b, and whose equivalent thickness (t_s) is determined as follows:

$$t_s = \sqrt[3]{\frac{12I_{sf}}{b_o}} \qquad (4.16)$$

where I_{sf} is the second moment of area of the full area of the multiple-stiffened element (including the intermediate stiffeners) about its own centroidal axis. The ratio b_o/t_s is then compared with the limiting value (w/t) to see if the equivalent element is fully effective. If b_o/t_s does not exceed the limiting w/t defined in Section B2.1, the entire multiple stiffened element shall be considered fully effective. However, if b_o/t_s exceeds the limiting w/t, the effective section properties shall be calculated on the basis of the true thickness (t) but with reductions in the width of each stiffened compression element and the area of each stiffener as specified in Section B5(c).

An improved method for multiple longitudinal intermediate stiffeners has been proposed by Schafer and Peköz (Ref. 4.13). It is based on independent computation of the local and distortional buckling stresses with an appropriate strength reduction factor for distortional buckling.

4.6 EXAMPLES

4.6.1 Hat Section in Bending

Problem

Determine the design positive bending moment for bending about a horizontal axis of the hat section in Figure 4.10. The yield stress of the material is 50 ksi. Assume the element lies along its centerline and eliminate thickness effects. The section numbers refer to the AISI Specification.

$D = 4.0$ in. $B = 9.0$ in. $t - 0.060$ in. $R = 0.125$ in.
$L_f = 0.75$ in. $B_f = 3.0$ in. $F_y = 50$ ksi $E = 29,500$ ksi

Solution

I. Properties of 90° corner (Elements 2 and 6)

Corner radius $\qquad r = R + \dfrac{t}{2} = 0.155$ in.

Length of arc $\qquad u = 1.57r = 0.243$ in.

Distance of centroid from $\qquad c = 0.637r = 0.099$ in.
center of radius

I' of corner about its own centroidal axis is negligible.

II. Nominal flexural strength (M_n) (Section C3.1.1)
Based on Initiation of Yielding [Section C3.1.1(a)]
A. Computation of I_x assuming web is fully effective
Assume $f = F_y$ in the top fibers of the section

Element 4: Webs

$h = D - 2(R + t) = 3.63$ in.

$\dfrac{h}{t} = 60.5 \qquad$ (this value must not exceed 200)

[Section B1.2(a)]

Element 5: Compression Flange

$w = B - 2(R + t) = 8.63$ in.

$\dfrac{w}{t} = 144 \qquad$ (this value must not exceed 500)

[Section B1.1(a)(2)]

Section B2.1(a)

$k = 4 \qquad \lambda = \dfrac{1.052}{\sqrt{k}} \left(\dfrac{w}{t}\right) \sqrt{\dfrac{f}{E}} = 3.169 \qquad$ (Eq. B2.1-4)

$\rho = \dfrac{1 - 0.22/\lambda}{\lambda} = 0.2984 \qquad$ (Eq. B2.1-3)

$b = \rho w = 2.575$ in. $\quad$ since $\lambda > 0.673$

(Eqs. B2.1-1 and B2.1-2)

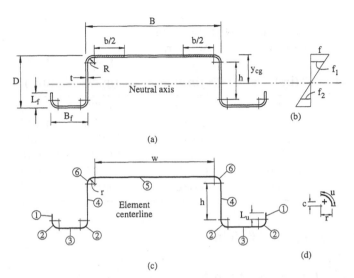

FIGURE 4.10 Hat sections in bending: (a) cross section; (b) stresses; (c) line element model; (d) radius.

Element 1: Lips

$$L_u = L_f - (R + t) = 0.565 \text{ in.}$$

Element 3: Bottom Flanges

$$b_f = B_f - 2(R + t) = 2.63 \text{ in.}$$

Calculate second moment of area of effective section

L_i is the effective length of all elements of type i, and y_i is the distance from the top fiber:

$$L_1 = 2L_u \quad y_1 = D - (R + t) - \frac{L_u}{2} \quad I'_1 = \frac{2L_u^3}{12} = 0.030 \text{ in.}^3$$

$$L_2 = 4u \quad y_2 = D - (R + t) + c$$

$$L_3 = 2b_f \quad y_3 = D - \frac{t}{2}$$

$$L_4 = 2h \quad y_4 = (R + t) + \frac{h}{2} \qquad I'_4 = \frac{2h^3}{12} = 7.972 \text{ in.}^3$$

$$L_5 = b \quad y_5 = \frac{t}{2}$$

$$L_6 = 2u \quad y_6 = (R + t) - c$$

Element no., i	Effective length, L_i (in.)	Distance from top fiber, y_i (in.)	$L_i y_i$ (in.^{2})	$L_i y_i^2$ (in.^{3})
1	1.130	3.532	3.992	14.101
2	0.973	3.914	3.809	14.910
3	5.260	3.970	20.882	82.902
4	7.260	2.000	14.520	29.040
5	2.575	0.030	0.077	0.002
6	0.487	0.086	0.042	0.004

$$\sum_{i=1}^{6} L_i = 17.685 \text{ in.}$$

$$\sum_{i=1}^{6} L_i y_i = 43.323 \text{ in.}^2$$

$$\sum_{i=1}^{6} L_i (y_i)^2 = 140.959 \text{ in.}^3$$

Determine distance (y_{cg}) of neutral axis from top fiber

$$y_{cg} = \frac{\sum_{i=1}^{6} L_i y_i}{\sum_{i=1}^{6} L_i} = 2.450 \text{ in.}$$

$$I_{Ly^2} = \sum_{i=1}^{6} L_i y_{cg}^2 = 106.127 \text{ in.}^3$$

$$I'_x = \sum_{i=1}^{6} L_i y_i^2 + I'_1 + I'_4 - I_{Ly^2} = 42.834 \text{ in.}^3$$

$$I_x = I'_x t = 2.570 \text{ in.}^4$$

B. Check Web (Section B2.3)

$$f_1 = [y_{cg} - (R+t)] \frac{f}{y_{cg}} = 46.22 \text{ ksi}$$

$$f_2 = -[h + (R+t) - y_{cg}] \frac{f}{y_{cg}} = -27.87 \text{ ksi}$$

$$\psi = \frac{f_2}{f_1} = -0.603 \qquad \text{(Eq. B2.3-5)}$$

$$k = 4 + 2(1 - \psi)^3 + 2(1 - \psi) = 15.442 \text{(Eq. B2.3-4)}$$

$$\lambda = \frac{1.052}{\sqrt{k}} \left(\frac{h}{t}\right)\sqrt{\frac{f_1}{E}} = 0.641 \qquad \text{(Eq. B2.1-4)}$$

$$b_e = b = h = 3.63 \text{ in.} \qquad \text{since } \lambda < 0.673$$

$$b_1 = \frac{b_e}{3 - \psi} = 1.008 \text{ in.} \qquad \qquad \text{Eq. B2.3-1)}$$

$$b_2 = \frac{b_e}{2} = 1.815 \text{ in.} \qquad \text{since } \psi < -0.236$$

$$\text{(Eqs. B2.3-2 and B2.3-3)}$$

$$b_{cw} = y_{cg} - (R + t) = 2.265 \text{ in.}$$

(b_{cw} is the compression portion of the web)

$$b_1 + b_2 = 2.823 \text{ in.}$$

Since $b_1 + b_2 > b_{cw}$, $b_e = 2.265$ in. Since $b_e = b_{cw}$, the web is fully effective and no iteration is required.

C. Calculation of effective section modulus (S_e) for compression flange stressed to yield

$$S_e = \frac{I_x}{y_{cg}} = 1.049 \text{ in.}^3$$

D. Calculation of nominal flexural strength (M_n) and design flexural strength (M_d)

$$M_n = S_e F_y = 52.457 \text{ kip-in.} \quad \text{(Eq. C3.1.1-1)}$$

LRFD $\quad \phi_b = 0.95$

$$M_d = \phi_b M_n = 49.834 \text{ kip-in.}$$

ASD $\quad \Omega_b = 1.67$

$$M_d = \frac{M_n}{\Omega_b} = 34.411 \text{ kip-in.}$$

4.6.2 Hat Section in Bending with Intermediate Stiffener in Compression Flange

Problem

Determine the design positive bending moment about a horizontal axis of the hat section in Figure 4.10 when an intermediate stiffener is added to the center of the compression flange as shown in Figure 4.11. The following section numbers refer to the AISI Specification:

$D = 4.0$ in. $B = 9.0$ in. $t = 0.060$ in. $R = 0.125$ in.

$L_f = 0.75$ in. $B_f = 3.0$ in. $d_s = 0.30$ in.

$F_y = 50$ ksi $E = 29,500$ ksi

Solution

I.   Properties of 90° corner (Elements 2 and 6)

Corner radius $\qquad\qquad\qquad r = R + \dfrac{t}{2} = 0.155$ in.

Length of arc $\qquad\qquad\quad u = 1.57r = 0.243$ in.

Distance of centroid from $\quad c = 0.637r = 0.099$ in.
center of radius

I' of corner about its own centroidal axis is negligible

II. Nominal flexural strength (M_n) (Section C3.1.1)

Based on Initiation of Yielding [Section C3.1.1(a)]

A.   Computation of I_x assuming web is fully effective

Assume $f = F_y$ in the top fibers of the section.

Element 4: Webs

$h = D - 2(R + t) = 3.63$ in.

$\dfrac{h}{t} = 60.5$ (this value must not exceed 200)

[Section B1.2(a)]

Compression Elements

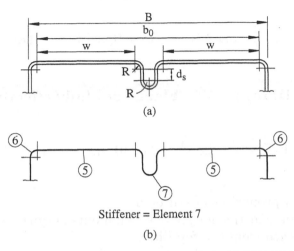

(a)

(b)

Stiffener = Element 7

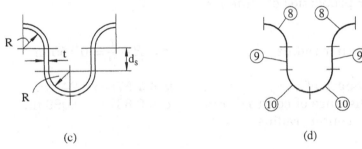

(c)   (d)

FIGURE 4.11   Hat section flange with intermediate stiffener: (a) cross section of compression flange; (b) line element model of flange; (c) stiffener section; (d) line element model of stiffener.

Element 5: Compression Flange

$$b_o = B - 2(R + t) = 8.63 \text{ in.}$$

$$\frac{b_o}{t} = 144 \qquad \text{(this value must not exceed 500)}$$

[Section B1.1(a)(2)]

Section B4.1(a) Uniformly compressed elements with one intermediate stiffener (see Figure 4.11)

$$S = 1.28\sqrt{\frac{E}{f}} = 31.091 \qquad 3S = 93.273 \qquad \text{(Eq. B4-1)}$$

Case III ($b_o/t > 3S$): Flange not fully effective

$$I_a = t^4\left(\frac{128b_o/t}{S} - 285\right) = 0.0040 \text{ in.}^4 \qquad \text{(Eq. B4.1-6)}$$

Full section properties of stiffener
L_i is the effective length of all elements of type i, and y_i is the distance from the top fiber.
Corner properties of stiffener

Corner radius $\qquad\qquad\qquad r = R + \dfrac{t}{2} = 0.155$ in.

Length of arc $\qquad\qquad\quad u = 1.57r = 0.243$ in.
Distance of centroid from $\quad c = 0.637r = 0.099$ in.
    center radius

$$L_8 = 2u \quad y_8 = (R + t) - c$$

$$L_9 = 2d_s \quad y_9 = (R + t) + \frac{d_s}{2} \qquad I'_9 = \frac{2d_s^3}{12} = 0.0045 \text{ in.}^3$$

$$L_{10} = 2u \quad y_{10} = (R + t) + d_s + c$$

Element no., i	Effective length, L_i (in.)	Distance from top fiber, y_i (in.)	L_iy_i (in.^{2})	$L_iy_i^2$ (in.^{3})
8	0.487	0.086	0.042	0.0036
9	0.600	0.335	0.201	0.067
10	0.487	0.584	0.284	0.166

$$\sum_{i=8}^{10} L_i = 1.573 \text{ in.}$$

$$\sum_{i=8}^{10} L_i y_i = 0.527 \text{ in.}^2$$

$$\sum_{i=8}^{10} L_i y_i^2 = 0.237 \text{ in.}^3$$

$$y_{cg} = \frac{\sum_{i=8}^{10} L_i y_i}{\sum_{i=8}^{10} L_i} = 0.335 \text{ in.}$$

$$I_{Ly^2} = \sum_{i=8}^{10} L_i y_{cg}^2 = 0.177 \text{ in.}^3$$

$$I_s' = \left[\sum_{i=8}^{10} L_i (y_i)^2 + I_9' \right] - I_{Ly^2} = 0.065 \text{ in.}^3$$

$$I_s = I_s' t = 0.0039 \text{ in.}^4$$

$$A_s' = \left(\sum_{i=8}^{10} L_i \right) t = 0.094; \text{ in.}^2$$

Reduced area of stiffener (A_s)

$$A_s = \left(\frac{I_s}{I_a} \right) A_s' = 0.092 \text{ in.}^2 \qquad \text{(Eq. B4.1-8)}$$

$$L_s = \frac{A_s}{t} = 1.535 \text{ in.}$$

Continuing with Element 5

$$k = 3 \left(\frac{I_s}{I_a} \right)^{1/3} + 1 = 3.98 \leq 4.00 \qquad \text{(Eq. B4.1-7)}$$

$$w = \frac{B - 2(R+t) - 2R - 2(R+t)}{2} = 4.005 \text{ in.}$$

$$\frac{w}{t} = 67$$

103

$$\lambda = \frac{1.052}{\sqrt{k}}\left(\frac{w}{t}\right)\sqrt{\frac{f}{E}} = 1.450 \qquad \text{(Eq. B2.1-4)}$$

$$\rho = \frac{1 - 0.22/\lambda}{\lambda} = 0.585 \qquad \text{(Eq. B2.1-3)}$$

$$b = \rho w = 2.343 \text{ in.} \qquad \text{since } \lambda > 0.673$$

Element 1: Lip

$$L_u = L_f - (R+t) = 0.565 \text{ in.}$$

Element 3: Flange

$$b_f = B_f - 2(R+t) = 2.63 \text{ in.}$$

L_i is the effective length of all elements of type i and y_i is the distance from the top fiber:

$$L_1 = 2L_u \quad y_1 = D - (R+t) - \frac{L_u}{2} \quad I'_1 = \frac{2L_u^3}{12} = 0.03 \text{ in.}^3$$

$$L_2 = 4u \quad y_2 = D - (R+t) + c$$

$$L_3 = 2b_f \quad y_3 = D - \frac{t}{2}$$

$$L_4 = 2h \quad y_4 = (R+t) + \frac{h}{2} \qquad I'_4 = \frac{2h^3}{12} = 7.972 \text{ in.}^3$$

$$L_5 = 2w \quad y_5 = \frac{t}{2}$$

$$L_6 = 2u \quad y_6 = (R+t) - c$$

$$L_7 = L_s \quad y_7 = (R+t) + \frac{d_s}{2} \qquad I'_7 = I'_s$$

Element no., i	Effective length, L_i (in.)	Distance from top fiber, y_i (in.)	$L_i y_i$ (in.^{2})	$L_i y_i^2$ (in.^{3})
1	1.130	3.533	3.992	14.101
2	0.973	3.914	3.810	14.910
3	5.260	3.970	20.882	82.902
4	7.260	2.000	14.520	29.040
5	4.686	0.030	0.141	4.218
6	0.487	0.086	0.042	3.622
7	1.535	0.335	0.514	0.173

$$\sum_{i=1}^{7} L_i = 21.331 \text{ in.}$$

$$\sum_{i=1}^{7} = L_i y_i = 43.9 \text{ in.}^2$$

$$\sum_{i=1}^{7} L_i y_i^2 = 141.133 \text{ in.}^3$$

Determine distance (y_{cg}) of neutral axis from top fiber:

$$y_{cg} = \frac{\sum_{i=1}^{7} L_i y_i}{\sum_{i=1}^{7} L_i} = 2.058 \text{ in.}$$

$$I_{Ly^2} = \sum_{i=1}^{7} L_i y_{cg}^2 = 90.348 \text{ in.}^3$$

$$I_x' = \sum_{i=1}^{7} L_i y_i^2 + I_1' + I_4' - I_{Ly^2} = 58.787 \text{ in.}^3$$

$$I_x = I_x' t = 3.527 \text{ in.}^4$$

B. Check Web (Section B2.3)

$$f_1 = [y_{cg} - (R+t)]\frac{f}{y_{cg}} = 45.51 \text{ ksi}$$

$$f_2 = -[h + (R+t) - y_{cg}]\frac{f}{y_{cg}} = -42.69 \text{ ksi}$$

$$\psi = \frac{f_2}{f_1} = -0.938 \qquad \text{(Eq. B2.3-5)}$$

$$k = 4 + 2(1 - \psi)^3 + 2(1 - \psi) = 22.434 \qquad \text{(Eq. B2.3-4)}$$

$$\lambda = \frac{1.052}{\sqrt{k}}\left(\frac{h}{t}\right)\sqrt{\frac{f_1}{E}} = 0.528 \qquad \text{(Eq. B2.1-4)}$$

$$b_e = 3.63 \text{ in.} \qquad \text{since } \lambda < 0.673$$

$$b_1 = \frac{b_e}{3 - \psi} = 0.922 \text{ in.} \qquad \text{(Eq. B2.3-1)}$$

$$b_2 = \frac{b_e}{2} = 1.815 \text{ in.} \qquad \text{since } \psi < -0.236$$

(Eqs. B2.3-2 and B2.3-3)

$$b_{cw} = y_{cg} - (R + t) = 1.873 \text{ in.}$$

(b_{cw} is the compression portion of the web)

$$b_1 + b_2 = 2.737 \text{ in.}$$

Since $b_1 + b_2 > b_{cw}$ then $b_e = 2.737$ in.

C. Calculation of effective section modulus (S_e) for compression flange stressed to yield

$$S_e = \frac{I_x}{y_{cg}} = 1.714 \text{ in.}^3$$

D. Calculation of nominal flexural strength (M_n) and design flexural strength (M_d)

$$M_n = S_e F_y = 85.69 \text{ kip-in.} \qquad \text{(Eq. C3.1.1-1)}$$

LRFD $\quad \phi_b = 0.95$

$$M_d = \phi_b M_n = 81.41 \text{ kip-in.}$$

ASD $\quad \Omega_b = 1.67$

$$M_d = \frac{M_n}{\Omega_b} = 51.31 \text{ kip-in.}$$

4.6.3 C-Section Purlin in Bending

Problem

Determine the design positive bending moment about the horizontal axis for the 8C0.060 purlin section shown in Figure 4.12. The yield stress of the material is 55 ksi. Assume the element lies along its centerline and eliminate thickness effects. The section modulus should be computed

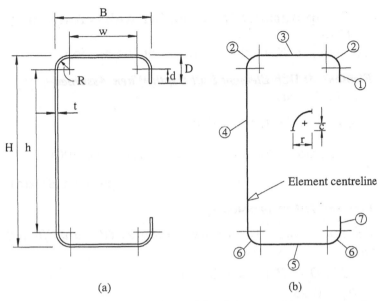

FIGURE 4.12 Purlin section with elements: (a) cross section; (b) line element model.

assuming the section is fully stressed ($f = F_y$). The following section numbers refer to the AISI Specification.

$H = 8.0$ in. $\qquad B = 2.75$ in. $\qquad t = 0.060$ in. $\qquad R = 0.1875$ in.
$D = 0.625$ in. $\qquad F_y = 55$ ksi $\qquad E = 29,500$ ksi

Solution

I. Properties of 90° corner (Elements 2 and 6)

Corner radius	$r = R + \frac{t}{2} = 0.218$ in.
length of arc	$u = 1.57r = 0.341$ in.
Distance of centroidal from center of radius	$c = 0.637r = 0.139$ in.
I' of corner about its own centroidal axis	$I_c' = 0.149r^3 = 0.0015$ in.^{3}

II. Nominal flexural strength (M_n) (Section C3.1.1)
Based on Initiation of Yielding [Section C3.1.1(a)]

107

A. Computation of I_x assuming web is fully effective

Assume $f = F_y$ in the top fibers of the section.

Element 4: Web Element Full Depth When Assumed Fully Effective

$h = H - 2(R + t) = 7.505$ in.

$\dfrac{h}{t} = 125.08$ (this value must not exceed 200)

[Section B1.2(a)]

Element 1: Compression Lip

Section B3.1 Uniformly compressed unstiffened element

$d = D - (R + t) = 0.378$ in.
$k_u = 0.43$

$$\lambda = \frac{1.052}{\sqrt{k_u}} \left(\frac{d}{t}\right)\sqrt{\frac{f}{E}} = 0.436 \qquad \text{(Eq. B2.1-4)}$$

$d'_s = d = 0.378$ in.   since $\lambda < 0.673$

$$I_s = \frac{d^3 t}{12} = 0.269 \times 10^{-3} \text{ in.}^4$$

f could be revised to f_3 in Figure 4.13, but this is not necessary in this case.

Element 3: Flange Flat

$w = B - 2(R + t) = 2.255$ in.

$\dfrac{w}{t} = 37.583$ (this value must not exceed 60)

[Clause B1.1(a)(1)]

Section B4.2 Uniformly compressed elements with an edge stiffener (see Figure 4.13)

$$S = 1.28\sqrt{\frac{E}{f}} = 29.644 \qquad \text{(Eq. B4-1)}$$

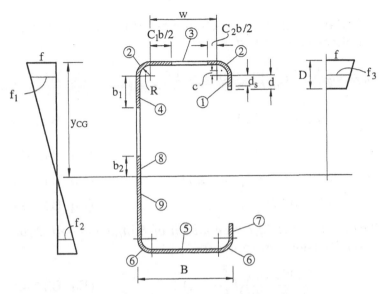

FIGURE 4.13 Bending stress with effective widths.

Case I ($w/t \leq S/3$) Flange fully effective without stiffener

Not applicable since $w/t > S/3$.

Case II ($S/3 < w/t < S$) Flange fully effective with $I_s \geq I_a$

$k_u = 0.43$

$$I_{a_2} = 399t^4\left(\frac{w/t}{S} - \sqrt{\frac{k_u}{4}}\right)^3 = 0.004265 \text{ in.}^4$$

$n_2 = 0.5$ \hfill (Eq. B4.2-4)

Case III ($w/t \geq S$) Flange not fully effective

$$I_{a_3} = t^4\left(115\frac{w/t}{S} + 5\right) = 0.001954 \text{ in.}^4 \qquad \text{(Eq. B4.2-11)}$$

$n_3 = 0.333$

Since $w/t \geq S$, then

$I_a = I_{a_3} = 0.001954 \text{ in.}^4$

$n = n_3 = 0.333$

Calculate buckling coefficients (k) and stiffener-reduced effective width (d_s):

$$k_a = 5.25 - 5\left(\frac{D}{w}\right) = 3.864 < 4.0 \qquad \text{(Eq. B4.2-8)}$$

$$C_2 = \frac{I_s}{I_a} = 0.138 < 1.0 \qquad \text{(Eq. B4.2-5)}$$

$$C_1 = 2 - C_2 = 1.862 \qquad \text{(Eq. B4.2-6)}$$

$$k = C_2^n(k_a - k_u) + k_u = 2.204 \qquad \text{(Eq. B4.2-7)}$$

$$d_s = c_2 d_s' = 0.052 \text{ in.} \qquad \text{(Eq. B4.2-9)}$$

Section B2.1(a) Effective width of flange element 3 for load capacity (see Figure 4.13)

$$\lambda = \frac{1.052}{\sqrt{k}}\left(\frac{w}{t}\right)\sqrt{\frac{f}{E}} = 1.15 \qquad \text{(Eq. B2.1-4)}$$

$$\rho = \frac{1 - 0.22/\lambda}{\lambda} = 0.703 \qquad \text{(Eq. B2.1-3)}$$

$$b_f = \rho w = 1.586 \text{ in.} \quad \text{since } \lambda > 0.673$$

Calculate second moment of area of effective section

L_i is the effective length of all elements of type i, and y_i is the distance from the top fiber:

$$L_1 = d_s \quad y_1 = (R + t) + \frac{d_s}{2} \qquad I_1' = \frac{d_s^3}{12} = 1.169 \times 10^{-5} \text{ in.}^3$$

$$L_2 = 2u \quad y_2 = (R + t) - c \qquad I_2' = 2I_c' = 3.066 \times 10^{-3} \text{ in.}^3$$

$$L_3 = b_f \quad y_3 = \frac{t}{2} \qquad I_3' = 0 \text{ in.}^3$$

$$L_4 = h \quad y_4 = (R + t) + \frac{h}{2} \qquad I_4' = \frac{h^3}{12} = 35.227 \text{ in.}^3$$

$$L_5 = w \quad y_5 = D - \frac{t}{2} \qquad I_5' = 0 \text{ in.}^2$$

$$L_6 = 2u \quad y_6 = D - (R + t) + c \qquad I_6' = 2I_c' = 3.066 \times 10^{-3} \text{ in.}^3$$

$$L_7 = d \quad y_7 = D - (R + t) - \frac{d}{2} \qquad I_7' = \frac{d^3}{12} = 4.483 \times 10^{-3} \text{ in.}^3$$

Element no., i	Effective length, L_i (in.)	Distance from top fiber, y_i (in.)	$L_i y_i$ (in.^{2})	$L_i y_i^2$ (in.^{3})
1	0.052	0.274	0.0142	0.004
2	0.683	0.109	0.0744	0.008
3	1.586	0.030	0.0476	0.001
4	7.505	4.000	30.0200	120.080
5	2.255	7.970	17.9723	143.240
6	0.683	7.891	5.3892	42.526
7	0.378	7.564	2.8553	21.597

$$\sum_{i=1}^{7} L_i = 13.141 \text{ in.} \qquad \sum_{i=1}^{7} L_i y_i = 56.373 \text{ in.}^2$$

$$\sum_{i=1}^{7} L_i y_i^2 = 327.456 \text{ in.}^3$$

Determine distance (y_{cg}) of neutral axis from top fiber:

$$y_{cg} = \frac{\sum_{i=1}^{7} L_i y_i}{\sum_{i=1}^{7} L_i} = 4.29 \text{ in.} \qquad I_{Ly^2} = \sum_{i=1}^{7} L_i y_{cg}^2 = 241.828 \text{ in.}^3$$

$$I_x' = \sum_{i=1}^{7} L_i y_i^2 + \sum_{i=1}^{7} I_i' - I_{Ly^2} = 120.865 \text{ in.}^3$$

$$I_x = I_x' t = 7.252 \text{ in.}^4 \qquad S_{ex} = \frac{I_x}{y_{cg}} = 1.691 \text{ in.}^3$$

B. Check Web (Section B2.3)

$$f_1 = [y_{cg} - (R+t)]\frac{f}{y_{cg}} = 51.827 \text{ ksi}$$

$$f_2 = -[h + (R+t) - y_{cg}]\frac{f}{y_{cg}} = -44.396 \text{ ksi}$$

$$\psi - \frac{f_2}{f_1} = -0.857 \qquad\qquad \text{(Eq. B2.3-5)}$$

$$k = 4 + 2(1 - \psi)^3 + 2(1 - \psi) = 20.513 \qquad \text{(Eq. B2.3-4)}$$

$$\lambda = \frac{1.052}{\sqrt{k}} \left(\frac{h}{t}\right) \sqrt{\frac{f_1}{E}} = 1.218 \qquad \text{(Eq. B2.1-4)}$$

$$\rho = \frac{1 - 0.22/\lambda}{\lambda} = 0.673 \qquad \text{(Eq. B2.1-3)}$$

$$b_e = \rho h = 5.05 \text{ in.} \qquad \text{since } \lambda > 0.673 \qquad \text{(Eq. B2.1-2)}$$

$$b_1 = \frac{b_e}{3 - \psi} = 1.309 \text{ in.} \qquad \text{(Eq. B2.3-1)}$$

$$b_2 = \frac{b_e}{2} = 2.525 \text{ in.} \qquad \text{since } \psi < -0.236 \quad \text{(Eq. B2.3-2)}$$

$$b_{cw} = y_{cg} - (R + t) = 4.042 \text{ in.}$$

$$b_1 + b_2 = 3.834 \text{ in.} < b_{cw} = 4.042 \text{ in.}$$

Web is not fully effective in this case since $b_1 + b_2 < b_{cw}$.

C. Iterate on web effective width

Try a value of the centroid position and iterate until convergence:

$$y_{cg} = 4.35 \text{ in.}$$

D. Check Web (Section B2.3)

$$f_1 = [\, y_{cg} - (R + t)\,]\frac{f}{y_{cg}} = 51.871 \text{ ksi}$$

$$f_2 = -[h + (R + t) - y_{cg}]\frac{f}{y_{cg}} = -43.02 \text{ ksi}$$

$$\psi = \frac{f_2}{f_1} = -0.829 \qquad \text{(Eq. B2.3-5)}$$

$$k = 4 + 2(1 - \psi)^3 + 2(1 - \psi) = 19.903 \qquad \text{(Eq. B2.3-4)}$$

$$\lambda = \frac{1.052}{\sqrt{k}} \left(\frac{h}{t}\right) \sqrt{\frac{f_1}{E}} = 1.237 \qquad \text{(Eq. B2.1-4)}$$

$$\rho = \frac{1 - 0.22/\lambda}{\lambda} = 0.665 \qquad \text{(Eq. B2.1-3)}$$

$$b_e = \rho h = 4.989 \text{ in.} \qquad \text{since } \lambda > 0.673 \qquad \text{(Eq. B2.1-2)}$$

$$b_1 = \frac{b_e}{3 - \psi} = 1.303 \text{ in.} \qquad \text{(Eq. B2.3-1)}$$

$$b_2 = \frac{b_e}{2} = 2.494 \text{ in.} \qquad \text{since } \psi < -0.236 \qquad \text{(Eq. B2.3-2)}$$

Break web into three elements as shown in Figure 4.13. Elements 4 and 8 are the top and bottom effective portions of the compression zone, and Element 9 is the web tension zone which is fully effective.

$$L_4 = b_1 \qquad y_4 = (R + t) + \frac{b_1}{2} \qquad I_4' = \frac{b_1^3}{12} = 0.184 \text{ in.}^3$$

$$L_8 = b_2 \qquad y_8 = y_{cg} - \frac{b_2}{2} \qquad I_8' = \frac{b_2^3}{12} = 1.293 \text{ in.}^3$$

$$L_9 = H - y_{cg} \qquad y_9 = H - (R + t) \qquad I_9' = \frac{L_9^3}{12} = 3.283 \text{ in.}^3$$

$$- (R + t) \qquad\qquad - \frac{L_9}{2}$$

Element no., i	Effective length, L_i (in.)	Distance from top fiber, y_i (in.)	$L_i y_i$ (in.^{2})	$L_i y_i^2$ (in.^{3})
1	0.052	0.274	0.014	0.003
2	0.683	0.109	0.074	0.008
3	1.586	0.030	0.048	0.001
4	1.303	0.899	1.171	1.053
5	2.255	7.970	17.972	143.240
6	0.683	7.891	5.389	42.526
7	0.378	7.564	2.855	21.597
8	2.494	3.103	7.740	24.014
9	3.403	6.051	20.589	124.591

$$\sum_{i=1}^{9} L_i = 12.836 \text{ in.} \qquad \sum_{i=1}^{9} L_i y_i = 55.853 \text{ in.}^2$$

$$\sum_{i=1}^{9} L_i y_i^2 = 357.035 \text{ in.}^3$$

Determine distance (y_{cg}) of neutral axis from top fiber:

$$y_{cg} = \frac{\sum_{i=1}^{9} L_i y_i}{\sum_{i=1}^{9} L_i} = 4.35 \text{ in.}$$

The computed centroid position (y_{cg}) matches with the assumed value:

$$I_{Ly^2} = \sum_{i=1}^{9} L_i y_{cg}^2 = 243.035 \text{ in.}^3$$

$$I_x' = \sum_{i=1}^{9} L_i y_i^2 + \sum_{i=1}^{9} I_i' - I_{Ly^2} = 118.77 \text{ in.}^3$$

$$I_x = I_x' t = 7.126 \text{ in.}^4$$

E. Calculation of effective section modulus (S_{ex}) for compression flange stressed to yield

$$S_{ex} = \frac{I_x}{y_{cg}} = 1.638 \text{ in.}^3$$

The resulting effective section modulus (S_{ex}) allows for the reduction in the effective area of both the web and flange due to their slenderness.

F. Calculation of nominal flexural strength (M_n) and design flexural strength (M_d)

$$M_n = S_e F_y = 90.09 \text{ kip-in.} \quad \text{(Eq. C3.1.1-1)}$$

LRFD $\qquad \phi_b = 0.95$

$$M_d = \phi_b M_n = 86.54 \text{ kip-in.}$$

ASD $\qquad \Omega_d = 1.67$

$$M_d = \frac{M_n}{\Omega_b} = 53.95 \text{ kip-in.}$$

114

4.6.4 Z-Section Purlin in Bending

Problem

Determine the design positive bending moment about the horizontal axis for the 8Z060 purlin section shown in Figure 4.14. The yield stress of the material is 55 ksi. Assume the element lies along its centerline and eliminate thickness effects. The section modulus should be computed assuming the section is fully stressed ($f = F_y$). The follow-

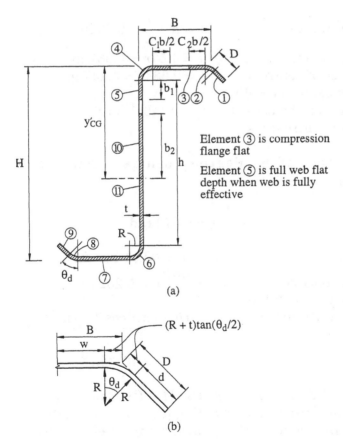

Element ③ is compression flange flat

Element ⑤ is full web flat depth when web is fully effective

(a)

(b)

FIGURE 4.14 Z-section with sloping lips. (a) effective cross section; (b) sloping lip and corner.

ing section and equation numbers refer to those in the AISI Specification.

$$H = 8.0 \text{ in.} \qquad B = 3.0 \text{ in.} \qquad \theta_d = 50°$$
$$t = 0.060 \text{ in.} \qquad R = 0.20 \text{ in.} \qquad D = 0.60 \text{ in.}$$
$$E = 29500 \text{ ksi} \qquad F_y = 55 \text{ ksi}$$

Solution

I.   Properties of 90° corners (Elements 4 and 6)

Corner radius $\qquad\qquad r = R + \dfrac{t}{2} = 0.23 \text{ in.}$

Length of arc $\qquad\qquad u = 1.57r = 0.361 \text{ in.}$

Distance of centroid $\qquad c = 0.637r = 0.147 \text{ in.}$
   from center of radius

I' of corner about its $\qquad I'_c = 0.149r^3 = 1.813 \times 10^{-3} \text{ in.}^3$
   own centroid xis

II.   Properties of θ_d degree corners (Elements 2 and 8)

$$\theta = \frac{\theta_d \pi}{180} \qquad\qquad\qquad \theta = 0.873 \text{ rad}$$

Length of arc $\qquad\qquad u_1 = r\theta = 0.201 \text{ in.}$

Distance of centroid $\qquad c_1 = \dfrac{\sin\theta}{\theta} r = 0.202 \text{ in.}$
   from center of radius

I' of corner about its own
   centroid axis $\qquad I'_{c_1} = \left(\dfrac{\theta + \sin\theta\cos\theta}{2} - \dfrac{\sin^2\theta}{\theta} \right) r^3$

$$= 1.227 \times 10^{-4} \text{ in.}^3$$

III. Nominal flexural strength (M_n) (Section C3.1.1)
Based on Initiation of Yielding [Section C3.1.1(a)]
A. Computation of I_x assuming web is fully effective
Assume $f = F_y$ in the top fibers of the section.

Element 5: Web Element Full Depth When Assumed Fully Effective

$$h = H - 2(R + t) = 7.48 \text{ in.}$$

$$\frac{h}{t} = 126.67 \quad \text{(this value must not exceed 200)}$$

[Section B1.2(a)]

Elements 1 and 9: Compression and Tension Flange Lips

Section B3.1 Uniformly compressed unstiffened elements

$$d = D - (R + t)\tan\frac{\theta}{2} = 0.479 \text{ in.}$$

$$k = 0.43 \qquad \text{[Section B3.1(a)]}$$

$$\lambda = \frac{1.052}{\sqrt{k}}\left(\frac{d}{t}\right)\sqrt{\frac{f}{E}} = 0.553 \qquad \text{(Eq. B2.1-4)}$$

$$d' = d = 0.479 \text{ in.} \quad \text{since } \lambda < 0.673$$

$$I_s = \frac{d^3 t}{12}\sin^2\theta = 3.22 \times 10^{-4} \text{ in.}^4$$

f could be revised to f_3 as shown for the channel section in Figure 4.13, but this is not necessary in this case.

Elements 3 and 7: Compression and Tension Flange Flats

$$w = B - (R + t) - (R + t)\tan\frac{\theta}{2} = 2.619 \text{ in.}$$

$$\frac{w}{t} = 44 \quad \text{(this value must not exceed 60)}$$

[Section B1.1(a)]

Section B4.2 Uniformly compressed elements with an edge stiffener

$$S - 1.28\sqrt{\frac{E}{f}} = 29.644 \qquad \text{(Eq. B4-1)}$$

117

Case I ($w/t \leq S/3$) Flange fully effective without stiffener

Not applicable ($w/t > S/3$)

Case II ($S/3 < w/t < S$) Flange fully effective with $I_s \geq I_a$

$k_u = 0.43$

$$I_{a_2} = 399\, t^4 \left(\frac{w/t}{S} - \sqrt{\frac{k_u}{4}} \right)^3 = 7.708 \times 10^{-3} \text{ in.}^4$$

(Eq. B4.2-4)

Case III ($w/t \geq S$) Flange not fully effective

$$I_{a_3} = t^4 \left(115 \frac{w/t}{S} + 5 \right) = 2.259 \times 10^{-3} \text{ in.}^4 \quad \text{(Eq. B4.2-11)}$$

$n_3 = 0.333$

Since $w/t \geq S$, then

$I_a = I_{a_3} = 2.259 \times 10^{-3} \text{ in.}^4$

$n = n_3 = 0.333$

Calculate buckling coefficient (k) and compression stiffener reduced effective width (d_s):

$$k_a = 5.25 - 5\left(\frac{D}{w} \right) = 4.10 > 4.0 \qquad \text{(Eq.B4.2-8)}$$

Hence use

$k_a = 4.0$

$$C_2 = \frac{I_s}{I_a} = 0.143 < 1.0 \qquad \text{(Eq. B4.2-5)}$$

$$C_1 = 2 - C_2 = 1.857 \qquad \text{(Eq. B4.2-6)}$$

$$k = C_2^n (k_a - k_u) + k_u = 2.296 \qquad \text{(Eq. B4.2-7)}$$

$$d_s = C_d d_s' = 0.068 \text{ in.} \qquad \text{(Eq. B4.2-9)}$$

Section B2.1 Effective width of flange element 3 for strength

$$\lambda = \frac{1.052}{\sqrt{k}}\left(\frac{w}{t}\right)\sqrt{\frac{f}{E}} = 1.308 \qquad \text{(Eq. B2.1-4)}$$

$$\rho = \frac{1 - 0.22/\lambda}{\lambda} = 0.636 \qquad \text{(Eq. B2.1-3)}$$

$$b = \rho w = 1.665 \text{ in.} \qquad \text{since } \lambda > 0.673 \qquad \text{(Eq. B2.1-2)}$$

Calculate second moment of area of effective section
L_i is the effective length of all elements of type i, and y_i is the distance from the top fiber:

$$L_1 = d_s \quad y_1 = \left(R + \frac{t}{2}\right)(1 - \cos\theta) \quad I_1' = \frac{d_s^3 \sin^2\theta}{12}$$

$$+ \frac{d_s}{2}\sin\theta + \frac{t}{2} \qquad = 1.554 \times 10^{-5} \text{ in.}^3$$

$$L_2 = u \quad y_2 = (R + t) - c_1 \qquad I_2' = I_{c1}' = 1.227 \times 10^{-4} \text{ in.}^3$$

$$L_3 = b \quad y_3 = \frac{t}{2} \qquad I_3' = 0 \text{ in.}^3$$

$$L_4 = u \quad y_4 = (R + t) - c \qquad I_4' = I_c' = 1.813 \times 10^{-3} \text{ in.}^3$$

$$L_5 = h \quad y_5 = (R + t) + \frac{h}{2} \qquad I_5' = \frac{h^3}{12} = 34.876 \text{ in.}^3$$

$$L_6 = u \quad y_6 = H - (R + t) + c \qquad I_6' = I_c' = 1.813 \times 10^{-3} \text{ in.}^3$$

$$L_7 = w \quad y_7 = H - \frac{t}{2} \qquad I_7' = 0 \text{ in.}^3$$

$$L_8 = u_1 \quad y_8 = H - (R + t) + c_1 \qquad I_8' = I_{c1}' = 1.227 \times 10^{-4} \text{ in.}^3$$

$$L_9 = d \quad y_9 = H - \left(\frac{R + t}{2}\right) \qquad I_9' = \frac{d^3}{12}\sin^2\theta\frac{-t}{2}$$

$$\times (1 - \cos\theta)\frac{-d}{2}\sin\theta\frac{-l}{2} \qquad = 5.366 \times 10^{-3} \text{ in.}^3$$

Element no., i	Effective length, L_i (in.)	Distance from top fiber, y_i (in.)	$L_i y_i$ (in.^{2})	$L_i y_i^2$ (in.^{3})
1	0.068	0.138	0.009	0.001
2	0.201	0.058	0.012	0.001
3	1.665	0.030	0.050	0.001
4	0.361	0.113	0.041	0.005
5	7.48	4.00	29.92	119.68
6	0.361	7.887	2.848	22.459
7	2.619	7.97	20.872	166.346
8	0.201	7.942	1.594	12.660
9	0.479	7.704	3.689	28.419

$$\sum_{i=1}^{9} L_i = 13.434 \text{ in.}$$

$$\sum_{i=1}^{9} L_i y_i = 59.034 \text{ in.}^2$$

$$\sum_{i=1}^{9} L_i y_i^2 = 349.572 \text{ in.}^3$$

$$\sum_{i=1}^{9} I_i' = 34.885 \text{ in.}^3$$

Determine distance (y_{cg}) of neutral axis from top fiber:

$$y_{cg} = \frac{\sum_{i=1}^{9} L_i y_i}{\sum_{i=1}^{9} L_i} = 4.394 \text{ in.}$$

$$I_{Ly^2} = \sum_{i=1}^{9} L_i y_{cg}^2 = 259.411 \text{ in.}^3$$

$$I_x' = \sum_{i=1}^{9} L_i y_i^2 + \sum_{i=1}^{9} I_i' - I_{Ly^2} = 125.046 \text{ in.}^3$$

$$I_x = I_x' t = 7.503 \text{ in.}^4$$

$$S_{ex} = \frac{I_x}{y_{cg}} = 1.707 \text{ in.}^3$$

B. Check Web (Section B2.3)

$$f_1 = [y_{cg} - (R+t)]\frac{f}{y_{cg}} = 51.75 \text{ ksi}$$

$$f_2 = -[h + (R+t) - y_{cg}]\frac{f}{y_{cg}} = -41.88 \text{ ksi}$$

$$\psi = \frac{f_2}{f_1} = -0.809 \qquad\qquad\qquad \text{(Eq. B2.3-5)}$$

$$k = 4 + 2(1 - \psi)^3 + 2(1 - \psi) = 19.464 \qquad \text{(Eq. B2.3-4)}$$

$$\lambda = \frac{1.052}{\sqrt{k}}\left(\frac{h}{t}\right)\sqrt{\frac{f_1}{E}} = 1.245 \qquad\qquad \text{(Eq. B2.1-4)}$$

$$\rho = \frac{1 - 0.22/\lambda}{\lambda} = 0.661 \qquad\qquad \text{(Eq. B2.1-3)}$$

$$b_e = \rho h = 4.946 \text{ in.} \qquad \text{since } \lambda > 0.673 \qquad \text{(Eq. B2.1-2)}$$

$$b_1 = \frac{b_e}{3 - \psi} = 1.298 \text{ in.} \qquad\qquad \text{(Eq. B2.3-1)}$$

$$b_2 = \frac{b_e}{2} = 2.473 \text{ in.} \qquad \text{since } \psi < -0.236 \quad \text{(Eq. B2.3-2)}$$

$$b_c = y_{cg} - (R+t) = 4.134 \text{ in.}$$

$$b_1 + b_2 = 3.772 \text{ in.} \; < b_c = 4.134 \text{ in.}$$

Web is not fully effective in this case since $b_1 + b_2 < b_c$.

C. Iterate on web effective width

Try a value of the centroid position and iterate until convergence:

$$y'_{cg} = 4.50 \text{ in.}$$

D. Check Web (Section B2.3)

$$f_1 = [y_{cg} - (R+t)]\frac{f}{y_{cg}} = 51.82 \text{ ksi}$$

$$f_2 = -[h + (R+t) - y_{cg}]\frac{f}{y_{cg}} = -39.6 \text{ ksi}$$

$$\psi = \frac{f_2}{f_1} = -0.764 \qquad\qquad\qquad \text{(Eq. B2.3-5)}$$

$$k = 4 + 2(1 - \psi)^3 + 2(1 - \psi) = 18.509 \qquad \text{(Eq. B2.3-4)}$$

$$\lambda = \frac{1.052}{\sqrt{k}} \left(\frac{h}{t} \right) \sqrt{\frac{f_1}{E}} = 1.278 \qquad \text{(Eq. B2.1-4)}$$

$$\rho = \frac{1 - 0.22/\lambda}{\lambda} = 0.648 \qquad \text{(Eq. B2.1-3)}$$

$$b_e = \rho h = 4.846 \text{ in.} \qquad \text{since } \lambda > 0.673 \qquad \text{(Eq. B2.1-2)}$$

$$b_1 = \frac{b_e}{3 - \psi} = 1.287 \text{ in.} \qquad \text{(Eq. B2.3-2)}$$

$$b_2 = \frac{b_e}{2} = 2.423 \text{ in.} \qquad \text{since } \psi < -0.236 \qquad \text{(Eq. B2.3-2)}$$

Break into three elements as shown in Figure 4.14. Elements 5 and 10 are the top and bottom effective portions of the compression zone, and Element 11 is the web tension zone which is fully effective

$$L_5 = b_1 \qquad y_5 = (R + t) + \frac{b_1}{2} \qquad I_5' = \frac{b_1^3}{12}$$
$$= 0.178 \text{ in.}^3$$

$$L_{10} = b_2 \qquad y_{10} = y_{cg} - \frac{b_2}{2} \qquad I_{10}' = \frac{b_2^3}{12}$$
$$= 1.186 \text{ in.}^3$$

$$L_{11} = H - y_{cg} \qquad y_{11} = H - (R + t) \qquad I_{11}' = \frac{(L_{11})^3}{12}$$
$$-(R + t) \qquad \qquad -\frac{L_{11}}{2} \qquad \qquad = 2.834 \text{ in.}^3$$

Element no., i	Effective length, L_i (in.)	Distance from top fiber, y_i (in.)	$L_i y_i$ (in.^{2})	$L_i y_i^2$ (in.^{3})
1	0.68	0.138	0.009	0.001
2	0.201	0.058	0.012	0.001
3	1.665	0.030	0.050	0.001

(continued)

Element no., i	Effective length, L_i (in.)	Distance from top fiber, y_i (in.)	$L_i y_i$ (in.^{2})	$L_i y_i^2$ (in.^{3})
4	0.361	0.113	0.041	0.005
5	1.287	0.904	1.164	1.052
6	0.361	7.887	2.848	22.459
7	2.619	7.97	20.872	166.346
8	0.201	7.942	1.594	12.660
9	0.479	7.704	3.689	28.419
10	2.423	3.288	7.968	26.203
11	3.240	6.120	19.829	121.352

$$\sum_{i=1}^{11} L_i = 12.905 \text{ in.}$$

$$\sum_{i=1}^{11} L_i y_i = 58.075 \text{ in.}^2$$

$$\sum_{i=1}^{11} L_i y_i^2 = 378.499 \text{ in.}^3$$

$$\sum_{i=1}^{11} I_i' = 4.207 \text{ in.}^3$$

Determine distance (y_{cg}) of neutral axis from top fiber:

$$y_{cg} = \frac{\sum_{i=1}^{11} L_i y_i}{\sum_{i=1}^{11} L_i} = 4.50 \text{ in.}$$

The computed centroid position (y_{cg}) matches with the assumed value:

$$I_{Ly^2} = \sum_{i=1}^{11} L_i y_{cg}^2 = 261.347 \text{ in.}^3$$

$$I_x' = \sum_{i=1}^{11} L_i y_i^2 + \sum_{i=1}^{11} I_i' - I_{Ly^2} = 121.359 \text{ in.}^3$$

$$I_x = I_x' t = 7.282 \text{ in.}^4$$

E. Calculation of effective section modulus (S_{ex} for compression flange stressed to yield

$$S_{ex} = \frac{I_x}{y_{cg}} = 1.618 \text{ in.}^3$$

The resulting effective section modulus (S_{ex}) allows for the reduction in the effective area of both the web and flange due to their slenderness and is the value when the section is yielded at the extreme compression fiber.

F. Calculation of nominal flexural strength (M_n) and design flexural strength (M_d)

$$M_n = S_e F_y = 88.99 \text{ kip-in.} \quad \text{(Eq. C3.1.1-1)}$$

LRFD $\phi_b = 0.95$

$$M_d = \phi_b M_n = 84.54 \text{ kip-in.}$$

ASD $\Omega_b = 1.67$

$$M_d = \frac{M_n}{\Omega_b} = 53.929 \text{ kip-in.}$$

REFERENCES

4.1 Bleich, F., Buckling Strength of Metal Structures, McGraw-Hill, New York, 1952.

4.2 Bulson, P. S., The Stability of Flat Plates, Chatto and Windus, London, 1970.

4.3 Allen, H. G. and Bulson, P., Background to Buckling, McGraw-Hill, New York, 1980.

4.4 Von Karman, T., Sechler, E. E., and Donnell, L. H., The strength of thin plates in compression, transactions ASME, Vol. 54, MP 54-5, 1932.

4.5 Winter, G., Strength of thin steel compression flanges, Transactions, ASCE, Vol. 112, Paper No. 2305, 1947, pp. 527–576.

4.6 Winter, G., Thin-Walled Structures—Theoretical Solutions and Test Results, Preliminary Publications of the Eighth Congress, IABSE, 1968, pp. 101–112.

4.7 Kalyanaraman, V., Peköz, T., and Winter, G., Unstiffened compression elements, Journal of the Structural Division, ASCE, Vol. 103, No. ST9, Sept. 1977, pp. 1833–1848.

4.8 Weng, C. C. and Peköz, T. B., Subultimate Behavior of Uniformly Compressed Stiffened Plate Elements, Research Report, Cornell University, Ithaca, NY, 1986.

4.9 Ortiz-Colberg, R. and Peköz, T. B., Load Carrying Capacity of Perforated Cold-Formed Steel Columns, Research Report No. 81–12, Cornell University, Ithaca, NY, 1981.

4.10 Peköz, T., Development of a Unified Approach to the Design of Cold-Formed Steel Members, American Iron and Steel Institute, Research Report CF87-1, March 1987.

4.11 Desmond, T. P., Peköz, T., and Winter, G., Edge stiffeners for thin-walled members, Journal of Structural Engineering, ASCE, 1981, 1-7(2), p. 329–353.

4.12 Desmond, T. P., Peköz, T., and Winter, G., Intermediate stiffeners for thin-walled members, Journal of Structural Engineering, ASCE, 1981, 107(4), pp. 627–648.

4.13 Schafer, B. and Peköz, T., Cold-formed steel members with multiple longitudinal intermediate stiffeners, Journal of Structural Engineering, ASCE, 1998, 124 (10), pp. 1175–1181.

5

Flexural Members

5.1 GENERAL

The design of flexural members for strength is usually
governed by one of the following limit states:

(a) Yielding of the most heavily loaded cross section
in bending including local and postlocal buckling
of the thin plates forming the cross section, or

(b) Elastic or inelastic buckling of the whole beam in a
torsional-flexural (commonly called lateral) mode,
or

(c) Yielding and/or buckling of the web subjected to
shear, or combined shear and bending, or

(d) Yielding and/or buckling of the web subjected to
bearing (web crippling) or combined bearing and
bending

The design for (a) determines the nominal section strength
(M_n) given by Eq. (5.1).

$$M_n = S_e F_y \qquad (5.1)$$

where S_e is the effective modulus about a given axis computed at the yield stress (F_y). Calculation of the effective section modulus at yield is described in Chapter 4. As specified in Section C3.1.1 of the AISI Specification, the capacity factor (ϕ_b) for computing the design section moment capacity is 0.95 for sections with stiffened or partially stiffened compression flanges and 0.90 for sections with unstiffened compression flanges.

The design for (b) determines the lateral buckling strength (M_n) given by Eq. (5.2).

$$M_n = S_c F_c \tag{5.2}$$

where

$$F_c = \frac{M_c}{S_f} \tag{5.3}$$

where M_c is the critical moment for lateral buckling, S_c is the effective section modulus for the extreme compression fiber computed at the critical stress (F_c), and S_f is the full unreduced section modulus for the extreme compression fiber. Equation (5.2) allows for the interaction effect of local buckling on lateral buckling. The use of the effective section modulus computed at the critical stress (F_c) rather than the yield stress (F_y) allows for the fact that the section may not be fully stressed when the critical moment is reached and, hence, the effective section modulus is not reduced to its value at yield. The method is called the unified approach and is described in detail in Ref. 4.10. Section C3.1.2 of the AISI Specification gives design rules for laterally unbraced and intermediately braced beams including I-beams, C- and Z-section beams when cross-sectional distortion does not occur. The basis of the design rules for lateral buckling is described in Section 5.2.

The basic behavior of C- and Z-sections is described in Section 5.3, and design methods for C- and Z-sections as part of metal building roof and wall systems are described in Chapter 10. These include the R-factor design approach in Sections C3.1.3 and C3.1.4 of the AISI Specification, which allows for the restraint from sheathing attached by screw fastening to one flange or by a standing seam roof system. Methods for bracing beams against lateral and torsional deformation are described in Section D3 of the AISI Specification and Sections 5.4 and 10.3 of this book. Allowance for inelastic reserve capacity of flexural members is included as Section C3.1.1(b) of the AISI Specification as an alternative to initial yielding described by Section C3.1.1(a) and specified by Eq. (5.1). Inelastic reserve capacity is described in Section 5.5 of this book.

The design for (c) requires computation of the nominal shear strength (V_n) of the beam and is fully described in Chapter 6. The design for (d) requires computation of the nominal web crippling strength (P_n) and is also described in Chapter 6. The interaction of both shear and web crippling with section moment is also described in Chapter 6.

5.2 TORSIONAL-FLEXURAL (LATERAL) BUCKLING

5.2.1 Elastic Buckling of Unbraced Simply Supported Beams

The elastic buckling moment (M_e) of a simply supported singly symmetric I-beam or T-beam bent about the x-axis perpendicular to the web as shown in Figure 5.1a with equal and opposite end moments and of unbraced length (L) is given in Refs 5.1 and 5.2 and is equal to

$$M_e = \frac{\pi\sqrt{EI_yGJ}}{L}\left[\frac{\pi\delta}{2} + \sqrt{\left(\frac{\pi\delta}{2}\right)^2 + \left(1 + \frac{\pi^2 EC_w}{GJL^2}\right)}\right] \quad (5.4)$$

where

$$\delta = \frac{\pm\beta_x}{L}\sqrt{\frac{EI_y}{GJ}} \quad (5.5)$$

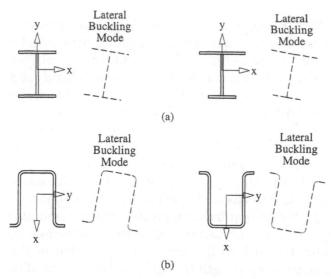

(a)

(b)

FIGURE 5.1 Lateral buckling modes and axes: (a) I and T-sections bent about x-axis; (b) hat and inverted hat sections bent about y axis.

and I_y, J and C_w are the minor axis second moment of area, the torsion constant, and warping constant, respectively. The value of δ is positive when the larger flange is in compression, is zero for doubly symmetric beams, and is negative when the larger flange is in tension.

The single-symmetry parameter (β_x) is a cross-sectional parameter defined by

$$\beta_x = \frac{\displaystyle\int_A (x^2 y + y^3)\, dA}{I_x} - 2y_0 \tag{5.6}$$

In the case of doubly symmetric beams, β_x is zero and Eq. (5.4) simplifies to

$$M_e = \frac{\pi\sqrt{EI_y GJ}}{L}\sqrt{1 + \frac{\pi^2 EC_w}{GJL^2}} \tag{5.7}$$

In the case of simply supported beams subjected to nonuniform moment, Eq. (5.7) can be modified by multiplying by

130

the factor C_b, which allows for the nonuniform distribution of bending moment in the beam:

$$M_e = \frac{C_b \pi \sqrt{EI_y GJ}}{L} \sqrt{1 + \frac{\pi^2 EC_w}{GJL^2}} \qquad (5.8)$$

The C_b factor is discussed in Section 5.2.2.

The elastic buckling moments (M_e) in Section C3.1.2 of the AISI Specification are expressed in terms of the elastic buckling stresses (σ_{ex}, σ_{ey}, σ_t). These stresses for an axially loaded compression member are fully described in Chapter 7. For example, Eq. [C3.1.2(6)] of the AISI Specification gives the elastic buckling moment for a singly symmetric section bent about the symmetry axis, or a doubly symmetric section bent about the x-axis,

$$M_e = C_b r_o A \sqrt{\sigma_{ey} \sigma_t} \qquad (5.9)$$

By substituting σ_{ey} and σ_t from Eqs. [C3.1.2(9)] and [C3.1.2(10)] of the AISI Specification, Eq. (5.9) becomes

$$M_e = C_b \frac{\pi \sqrt{EI_y GJ}}{K_y L_y} \sqrt{1 + \frac{\pi^2 EC_w}{GJ(K_t L_t)^2}} \qquad (5.10)$$

This is the same as Eq. (5.8) except that the unbraced length (L) is replaced by $K_y L_y$ for the effective length about the y-axis, and $K_t L_t$ for the effective length for twisting, as appropriate. Hence, Eq. [C3.1.2(6)] in Section C3.1.2 of the AISI Specification is a more general version of Eq. (5.8).

As a second example, Eq. [C3.1.2(7)] of the AISI Specification gives M_e for a singly symmetric section bent about the centroidal y-axis perpendicular to the symmetry x-axis, such as for the hat sections in Figure 5.1b, as

$$M_e = C_s A \sigma_{ex} \frac{j + C_s \sqrt{j^2 + r_o^2 (\sigma_t / \sigma_{ex})}}{C_{TF}} \qquad (5.11)$$

where j equals $\beta_y / 2$ and β_y is given by Eq. (5.6) with x and y interchanged, and C_s is $+1$ or -1 depending on whether the moment causes compression or tension on the shear center side of the centroid. Substituting σ_{ex} and σ_t from Eqs. [C3.1.2(8)] and [C3.1.2(10)] of the AISI Specification gives for Eq. (5.11)

$$M_e = \frac{\pi\sqrt{EI_x GJ}}{K_x L_x}\left[\frac{\pi\delta/2 + \sqrt{(\pi\delta/2)^2 + (1 + \pi^2 EC_w/GJ(K_t L_t)^2)}}{C_{TF}}\right]$$

(5.12)

where

$$\delta = \frac{\pm\beta_y}{K_x L_x}\sqrt{\frac{EI_x}{GJ}}$$

(5.13)

This is the same as Eqs. (5.4) and (5.5) except that

(i)   The x-axis is the axis of symmetry and the beam is bent about the y-axis.

(ii)   The C_{TF} factor to allow for nonuniform moment is included as in Eq. (5.14).

(iii) The unbraced length (L) is replaced by $K_x L_x$ for the effective length about the x-axis, and $K_t L_t$ for the effective length for twisting, as appropriate.

Hence, Eq. [C3.1.2(7)] in Section C3.1.2 of the AISI Specification is a more general version of Eqs. (5.4) and (5.5) with the x- and y-axes interchanged.

For a beam subjected to a clockwise moment (M_1) at the left-hand end and a clockwise moment (M_2) at the right-hand end, as shown in Figure 5.2a, a simple approximation for C_{TF}, as given in Refs. 5.1 and 8.1, is

$$C_{TF} = 0.6 - 0.4\left(\frac{M_1}{M_2}\right)$$

(5.14)

In Section C3.1.2 of the AISI Specification, a specific equation [C3.1.2(16)] for the elastic buckling moment of a point-symmetric Z-section is

$$M_e = \frac{\pi^2 EC_b dI_{yc}}{2L^2}$$

(5.15)

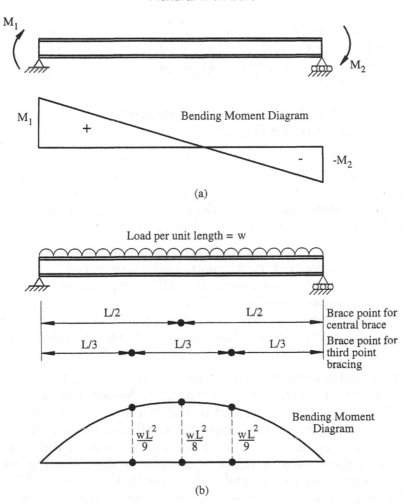

FIGURE 5.2 Simply supported beams: (a) beam under moment gradient; (b) beam under uniformly distributed load.

where I_{yc} is the second moment of area of the compression portion of the section about the centroidal axis of the full section parallel to the web, using the full unreduced section. This equation can be derived from Eq. (5.8) by putting $J = 0$, $C_w = I_y d^2/4$, and $I_{yc} = I_y/2$, and including an

133

additional factor $\frac{1}{2}$ to allow for the fact that a Z-section has an inclined principal axis, whereas I_{yc} is computed about the centroidal axis parallel to the web.

5.2.2 Continuous Beams and Braced Simply Suppported Beams

In practice, beams are not usually subjected to uniform moment or a linear moment distribution and are not always restrained by simple supports. Hence, if an accurate analysis of torsional-flexural buckling is to be performed, the following effects should be included:

(a) Type of beam support, including simply supported, continuous, and cantilevered
(b) Loading position, including top flange, shear center, and bottom flange
(c) Positioning and type of lateral braces
(d) Restraint provided by sheathing, including the membrane, shear, and flexural stiffnesses

A method of finite element analysis of the torsional-flexural buckling of continuously restrained beams and beam-columns has been described in Ref. 5.4 and was applied to the buckling of simply supported purlins with diaphragm restraints in Ref. 5.5 and continuous purlins in Ref. 5.3. The element used in these references is shown in Figure 5.3a and shown subjected to loading in Figure 5.3b. The loading allows for a uniformly distributed load located a distance a below the shear center.

A computer program PRFELB has been developed at the University of Sydney to perform a torsional-flexural buckling analysis of beam-columns and plane frames using the theory described in Refs. 5.2 and 5.4. The detailed method of operation of the program is described in Ref. 5.6.

The method has been applied to the buckling of simply supported beams subjected to uniformly distributed loads as shown in Figure 5.2b to determine suitable C_b factors for

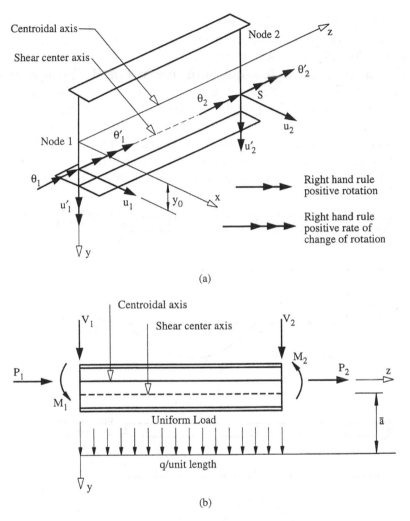

(a)

(b)

FIGURE 5.3 Element displacements and actions: (a) nodal displacements producing out-of-plane deformation; (b) element actions in the plane of the structure.

use with Eqs. (5.8) and (5.9). The loading was located at the tension flange, shear center axis, and compression flange. Three different bracing configurations were used, ranging from no intermediate bracing through central bracing to third-point bracing. Each brace, including the end supports, was assumed to prevent both lateral and torsional deformation. The element subdivisions used in the analysis are shown in Figure 5.4 for both central and third-point bracing. The resulting lateral deflections of the shear center in the buckling mode are also shown in Figure 5.4.

The computed buckling loads were then converted to buckling moments which were substituted into Eq. (5.8) to compute C_b values for each configuration analyzed. The resulting C_b values are summarized in Table 5.1. The term

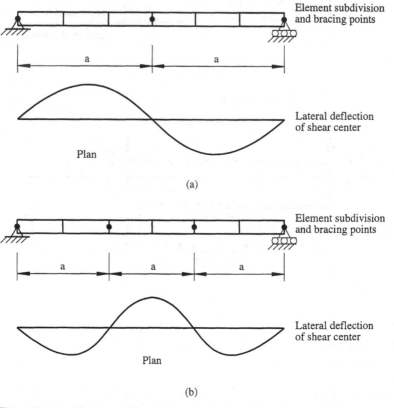

FIGURE 5.4 Intermediately braced simply supported beams: (a) buckling mode—central brace; (b) buckling mode—third-point bracing.

TABLE 5.1 C_b Factors for Simply Supported Beams Uniformly Loaded Within the Span

Loading position	No bracing $a = L$	One central brace $a = 0.5L$	Third point bracing $a = 0.33L$
Tension flange	1.92	1.59	1.47
Shear center	1.22	1.37	1.37
Compression flange	0.77	1.19	1.28

a in Table 5.1 is the bracing interval which replaces the length (L) in Eq. (5.8).

The C_b factors become larger as the loading moves from the compression flange to the tension flange since the buckling load and, hence, buckling moment are increased.

For the case of central bracing, the two half-spans buckle equally as shown in Figure 5.4a, and hence no elastic restraint is provided by one on the other. However, for the case of third-point bracing, the C_b factor has been computed for the central section between braces which is elastically restrained by the end sections, which are more lightly loaded, as shown in Figure 5.2b. If the C_b factor was computed for the central section, not accounting for the restraint provided by the end sections, then the C_b factors computed would be much closer to 1.0, since the central section is subjected to nearly uniform bending as shown in Figure 5.2b.

In addition to the restraint provided by braces, restraint is also provided by the sheathing through-fastened to one flange. The restraint provided by the sheathing is shown in Figure 5.5. If the sheathing has an adequate diaphragm (shear) stiffness (k_{ry}) in its plane, as defined in Figures 5.5a and 5.5b, it prevents lateral displacement of the flange to which it is connected. Further, if the sheathing has an adequate flexural stiffness (k_{rs}) as defined in Figure 5.5c, and is effectively connected to the purlin, it

137

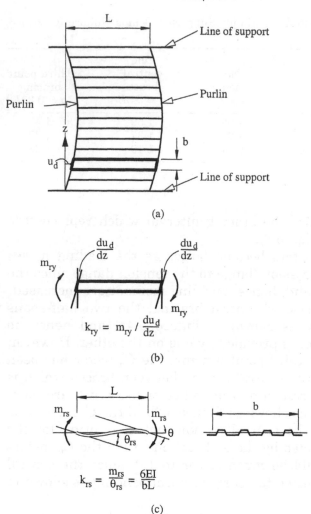

FIGURE 5.5 Sheathing stiffness: (a) plan of sheathing; (b) sheathing shear stiffness (k_{ry}); (c) sheathing flexural stiffness (k_{rs}).

can provide torsional restraint to the purlin. These stiffnesses can be incorporated in the finite element analysis described above to simulate the effect of sheathing.

A finite element analysis has been performed for the four-span continuous lapped purlin shown in Figure 5.6a

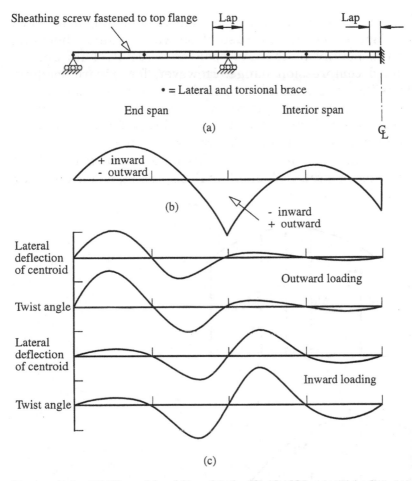

FIGURE 5.6   BMD and buckling modes for half-beam: (a) element subdivision; (b) BMD; (c) buckling modes.

by treating it as a doubly symmetric beam. It is restrained by bracing as well as sheathing on the top flange. The sheathing was assumed to provide a diaphragm shear stiffness of the type shown in Figure 5.5b, but no flexural stiffness of the type shown in Figure 5.5c. The loading was located at the level of the sheathing. The bending moment diagram is shown in Figure 5.6b, and the resulting buckling modes for both inward and outward loading are shown in Figure 5.6c. In the case of outward loading, buckling occurs mainly in the end span as a result of its unrestrained compression flange. However, for inward loading, buckling occurs mainly at the first interior support where there is a large negative moment and an unrestrained compression flange. The buckling loads for both inward and outward loading are approximately the same.

To analyze problems of this type in practice without recourse to a finite element analysis, the C_b factor has been

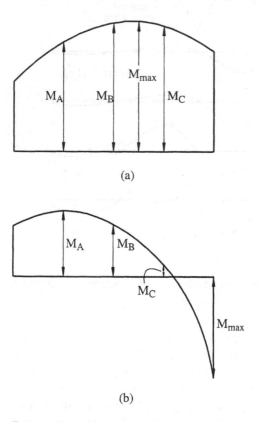

(a)

(b)

FIGURE 5.7 Bending moment distributions: (a) positive moment (or negative) alone; (b) positive and negative moments.

included in Section C3.1.2 as in Eqs. (5.8) and (5.15). The C_b factor is

$$C_b = \frac{12.5M_{\max}}{2.5M_{\max} + 3M_A + 4M_B + 3M_C} \tag{5.16}$$

where $M_{\max}$, M_A, M_B, and M_C are the *absolute values* of moment shown in Figures 5.7a and b. Equation (5.16) is Equation (C3.1.2-11) of the AISI Specification and was developed by Kirby and Nethercot (Ref. 5.7).

5.2.3 Bending Strength Design Equations

The interaction of yielding and lateral buckling has not been thoroughly investigated for cold-formed sections. Consequently, the design approach adopted in the AISI Specification is based on the Johnson parabola but factored by 10/9 to reflect partial plastification of the section in bending (Ref. 5.8).

The set of beam design equations in Section C3.1.2 of the AISI Specification is

$$M_c = \begin{cases} M_y & \text{for } M_e \geq 2.78M_y & (5.17) \\[2ex] 1.11M_y\left(1 - \dfrac{10M_y}{36M_e}\right) & \\ & \text{for } 2.78M_y > M_e > 0.56M_y & (5.18) \\[2ex] M_e & \text{for } M_e \leq 0.56M_y & (5.19) \end{cases}$$

They have been confirmed by research on beam-columns in Refs 5.9 and 5.10. The Eurocode 3 Part 1.3 (Ref. 1.7) beam design curve is also shown in Figure 5.8 for comparison. A detailed discussion of the lateral buckling strengths of unsheathed cold-formed beams is given in Ref. 5.11.

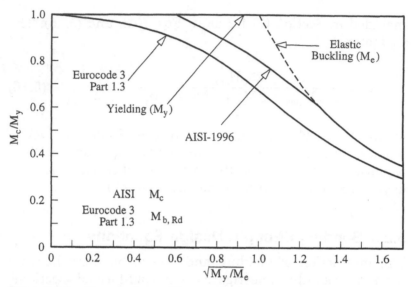

FIGURE 5.8   Comparison of beam design curves.

5.3 BASIC BEHAVIOR OF C- AND Z-SECTION FLEXURAL MEMBERS

5.3.1 Linear Response of C- and Z-Sections

5.3.1.1 General

Two basic section shapes are normally used for flexural members. They are Z-sections, which are point-symmetric, and C-sections, which are singly symmetric. For wind uplift, loading is usually perpendicular to the sheathing and parallel with the webs of the C- and Z-sections. If there is no lateral or torsional restraint from the sheathing, Z-sections will move vertically and deflect laterally as a result of the inclined principal axes, while C-sections will move vertically and twist as a result of the eccentricity of the load from the shear center. If there is full lateral and torsional restraint, both C- and Z-sections will bend in a plane parallel with their webs, and simple bending theory based on the section modulus computed for an axis perpendicular to the web can be used to compute the bending stress and shear flow distributions in the section. If the sections have the same thickness, flange width, web depth, and lip width, then the shear flow distributions are approximately the same for both sections.

5.3.1.2 Sections with Lateral Restraint Only

In order to investigate the basic linear behavior of C- and Z-sections when full lateral restraint is applied at the top flange with no torsional restraint, simply supported C- and Z-sections subjected to uplift along the line of the web and laterally restrained at the top flange have been analyzed with a linear elastic matrix displacement analysis which includes thin-walled tension (Refs. 5.12 and 5.13). The models used in the analysis are shown in Figure 5.9, where the sections have been subdivided into 20 elements and a lateral restraint is provided at each of the 19 internal nodes at the points where the uplift forces are applied. The uplift forces are statically equivalent to a uniformly distributed force (w) of 113 lb/in. The longitudinal stress distributions at the central cross sections are shown in Figure 5.10 and are almost identical for C- and Z-sections. The stress distributions show that the top flange remains in uniform stress, whereas the bottom flange undergoes transverse bending in addition to compression from the vertical component of bending. If plain C- and Z-sections were compared, the stress distributions would be identical. The lateral displacement at point C on the bottom flange of the sections in Figure 5.9 was computed to be 2.5 in. for the C-section and 2.4 in. for the Z-section. It can be concluded that C- and Z-sections with lateral restraint only from sheathing attached to one flange tend to behave similarly, whereas C- and Z-sections without restraint behave differently.

143

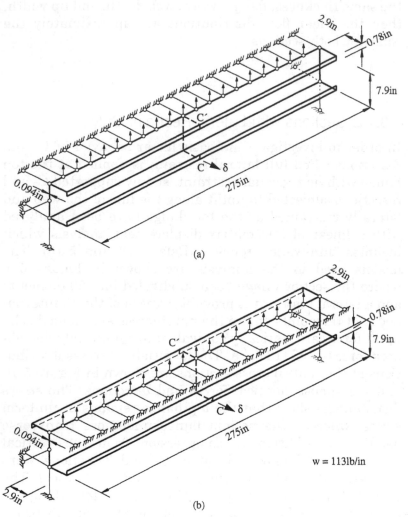

FIGURE 5.9 Simply supported beam models: (a) C-section; (b) Z-section.

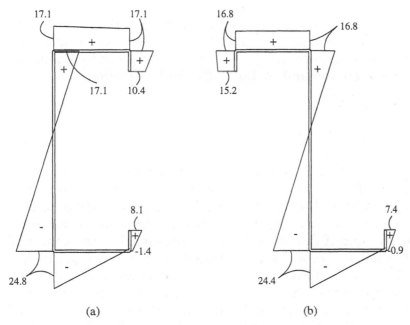

FIGURE 5.10 Longitudinal stress distributions at central cross section of beam models: (a) C-section, (b) Z-section.

5.3.1.3 Sections with Lateral and Torsional Restraint

The magnitude of the torsional restraint will govern the degree to which the unrestrained flange moves laterally for both C- and Z-sections. For C- and Z-sections under the wind uplift and restrained both laterally and torsionally by sheathing, deformation will tend to be as shown in Figure 5.11a. The forces per unit length acting on the purlins as shown in Figure 5.11b are q, representing uplift, p, representing prying, and r, representing lateral restraint at the top flange. The prying forces can be converted into a statically equivalent torque $(pb_f/2)$ as shown in Figure 5.11c. If the line of action of the uplift force (q) and lateral restraint force (r) are moved to act through the shear center for both sections, then a statically equivalent torque (m_t) results, as shown in Figure 5.11d, and is given by Eq. (5.20) for a Z-section and by Eq. (5.21) for a C-section:

$$m_t = \frac{pb_f}{2} - \frac{rb_w}{2} + \frac{qb_f}{2} \tag{5.20}$$

$$m_t = \frac{pb_f}{2} + \frac{rb_w}{2} - \frac{qb_f}{2} - qd_s \tag{5.21}$$

The $pb_f/2$ term is the torque resulting from the prying force. The $rb_w/2$ term is the torque resulting from the eccentricity of the restraining force from the shear center. The $qb_f/2$ term is the torque resulting from the eccentricity of the line of the screw fastener, presumed at the center of the flange, from the web centerline. The term qd_s in Eq. (5.21) is the torque resulting from the eccentricity of the shear center of the C-section from the web centerline. Since the forces in Figure 5.11d act through the shear center of each section, the value of m_t must be zero if the section does not twist. Hence, the different components of Eqs. (5.20) and (5.21) must produce a zero net result for m_t for untwisted sections.

For the Z-section, the value of r in Eq. (5.20) is given by Eq. (5.22) for a section which is untwisted and braced laterally:

$$r = \frac{-I_{xy}q}{I_x} \tag{5.22}$$

Using Eqs. (5.20) and (5.22) and assuming that the value of $-I_{xy}/I_x$ is equal to b_f/b_w, where the x- and y-axes are as shown in Figure 5.11b, so that the sign of I_{xy} is negative, then the second and third terms in Eq. (5.20) add to zero. Hence, the first term in Eq. (5.20) (prying force term) is zero if m_t is zero. Consequently, since it can be shown that the second and third terms in Eq. (5.20) cancel approximately for conventional Z-sections with the screw fastener in the center of the flange, the prying force will be small for Z-sections before structural instability of the unrestrained flange occurs.

Flexural Members

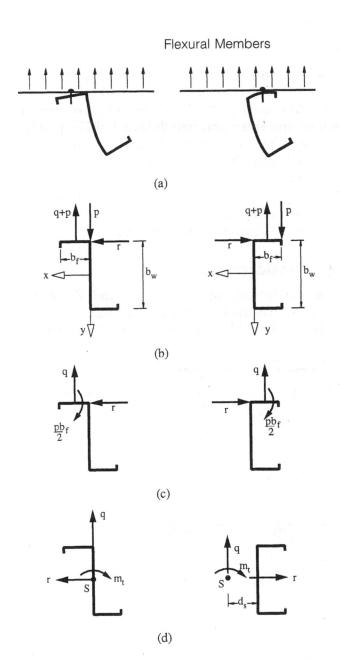

(a)

(b)

(c)

(d)

FIGURE 5.11 C- and Z-section statics: (a) deformation under uplift; (b) forces acting on purlins (a); (c) static equivalent of (b); (d) static equivalent of (c) with forces through shear center (S).

147

By comparison, the value of r in Eq. (5.21) for a C-section is zero if the section remains untwisted. Hence, for m_t to be zero, the first term in Eq. (5.21) (prying force term) must balance with the third and fourth terms in Eq. (5.21). That is, the torque from the prying forces is required to counterbalance the torque from the eccentricity of the uplift force in the center of the flange from the shear center. Consequently, the prying forces are much greater for C-sections, than they are for Z-sections before structural instability of the unrestrained flange occurs.

5.3.2 Stability Considerations

When studying the lateral stability of C- and Z-section flexural members restrained by sheathing, two basically different models of the buckling mode have been developed.

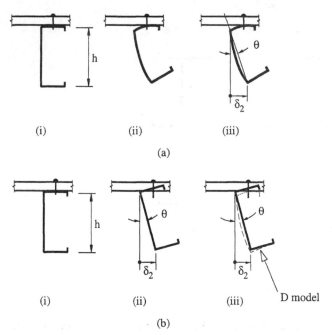

FIGURE 5.12 C-section twisting deformations: (a) distorted (D) section model; (b) undistorted (U) section model.

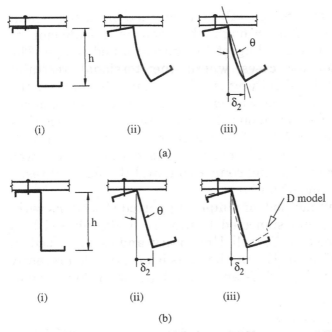

FIGURE 5.13 Z-section twisting deformations: (a) distorted (D) section model; (b) undistorted (U) section model.

These are the distorted section model (D model) shown in Figures 5.12a and 5.13a for C- and Z-sections, respectively, and the undistorted section model (U model) shown in Figures 5.12b and 5.13b for C- and Z-sections, respectively. The D model is similar to the behavior of sheeted purlins as observed in tests. The U model is based on the well-known thin-walled beam theories of Timoshenko (Ref. 3.6) and Vlasov (Ref. 5.14), so the stability can be computed by using the conventional stability theory of thin-walled beams with undistorted cross sections.

The D model has required the development of a new stability model to account for the section distortion which is not accounted for in the Timoshenko and Vlasov theories. In the D model, the unrestrained flange is usually in compression for purlins subjected to wind uplift. Hence, the compression flange will become unstable if the lateral restraint provided through the web is insufficient to prevent column buckling of the unrestrained flange. The usual model for cases where the torsional restraint provided by the sheathing is substantial is to consider the unrestrained compression flange as a compression member supported by a continuous elastic restraint of stiffness K as shown in Figure 5.14b. The elastic restraint of stiffness K shown in Figure 5.14b consists of the stiffness of the web and the purlin-sheeting connection, and the flexural resistance of the sheathing.

A design method for simply supported C-sections without bridging and subjected to uniformly distributed load was developed at Cornell University and is described in detail in Ref. 5.15. The method has been extended recently to continuous lapped C- and Z-sections, with and without

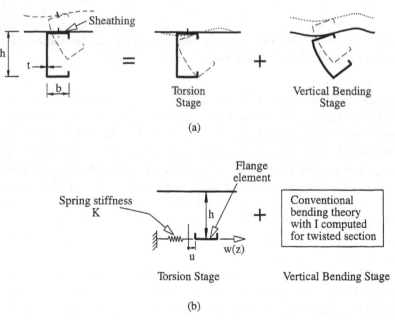

FIGURE 5.14 D model of C-section under uplift: (a) deflection; (b) models.

bridging, by Rousch and Hancock (Refs. 5.16–5.18). A brief summary of the method follows.

As shown in Figure 5.14a, the deflected configuration is assumed to be the sum of a torsion stage, which includes distortion of the section as a result of flexure of the web, and a vertical bending stage. It is assumed that the torsion stage is resisted by a flange element restrained by an elastic spring of stiffness K as shown in Figure 5.14b.

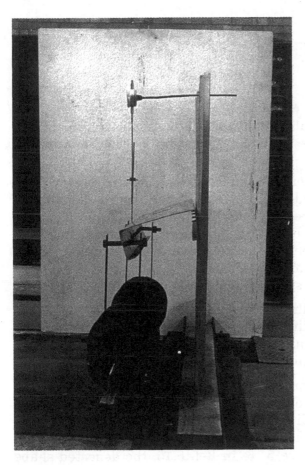

FIGURE 5.15 Test of C-section–sheathing combination to determine spring stiffness (K).

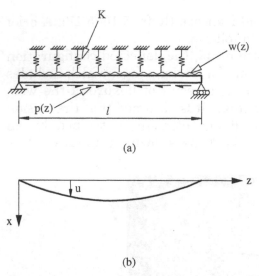

(a)

(b)

FIGURE 5.16 Idealized beam-column in D model: (a) forces and restraint on flange element; (b) plan of assumed deflected shape.

The flange element is assumed to be subjected to a distributed lateral load $w(z)$ per unit length and a distributed axial force $p(z)$ per unit length as shown in Figure 5.16a. It deflects u in its own plane as shown in Figure 5.16b. The torsional rotation, and hence the torsional stiffness (GJ_f), of the flange element is ignored.

The spring stiffness K depends upon the

(a) Flexibility of the web
(b) Flexibility of the connection between the tension flange and the sheathing
(c) Flexibility of the sheathing

A test of a short length of purlin through-fastened to sheathing as shown in Figure 5.15 is required to assess this total flexibility and, hence, K.

The flange element is then analyzed as a beam-column as shown in Figure 5.16. In Ref. 5.15, the deflected shape (u) of the beam-column is assumed to be a sine curve, and an energy analysis of the system is performed, including the flexural strain energy of the flange element, the strain energy of the elastic spring, the potential energy of the lateral loads [$w(z)$], and the potential energy of the axial loads [$p(z)$]. In Refs 5.16–5.18, a finite element beam-column model is used to determine the deflected shape (u) of the unrestrained flange.

The failure criteria involve comparing the maximum stress at the flange-web junction with the local buckling failure stress (F_{bw}), and the maximum stress at the flange lip junction with the distortional buckling failure stress.

5.4 BRACING

As described in Section 1.7.4, bracing of beams can be used to increase the torsional-flexural buckling load. In addition, non-doubly-symmetric sections, such as C- and Z-sections, twist or deflect laterally under load as a consequence of the loading which is not located through the shear center for C-sections or not in a principal plane for Z-sections. Consequently, bracing can be used to minimize lateral and torsional deformations and to transmit forces and torques to supporting members.

For C- and Z-sections used as beams, two basic situations exist as specified in Section D.3.2 of the AISI Specification.

(a) When the top flange is connected to deck or sheathing in such a way that the sheathing effectively restrains lateral deflection of the connected flange as described in Section D3.2.1

(b) When neither flange is so connected and bracing members are used to support the member as described in Section D3.2.2

A full discussion of bracing requirements for metal roof and wall systems is given in Section 10.4.

The formulae for the design of braces specified in Section D3.2.2 are based on an analysis of the torques and lateral forces induced in a C- or Z-section, respectively, assuming that the braces completely resist these forces. For the simple case of a concentrated load acting along the line of the web of a C-section as shown in Figure 5.17a, the brace connected to the C-section must resist a torque equal to the concentrated force times its distance from the shear center. For the case of a vertical force acting along the line

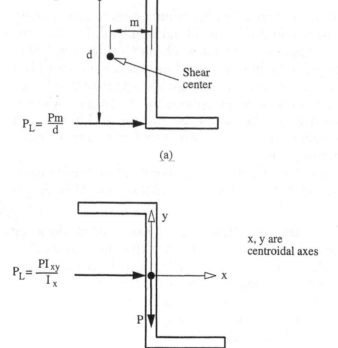

FIGURE 5.17 Theoretical bracing forces at a point of concentrated load: (a) C-section forces; (b) Z-section forces.

of the web of a Z-section as shown in Figure 5.17b, the brace must resist the load tending to cause the section to deflect perpendicular to the web. Section D3.2.2 also specifies how the bracing forces should be calculated when the loads are not located at the brace points or are uniformly distributed loads.

5.5 INELASTIC RESERVE CAPACITY

5.5.1 Sections with Flat Elements

As described in Section 1.7.10, a set of rules to allow for the inelastic reserve capacity of flexural members is included in Section C3.1.1(b) of the AISI Specification. These rules are based on the theory and tests in Ref. 5.19.

Cold-formed sections are generally not compact as are the majority of hot-rolled sections and hence are not capable of participating in plastic design. However, the neutral axis of cold-formed sections may not be at mid-depth, and initial yielding may take place in the tension flange. Also, since the web and flange thicknesses are the same for cold-formed sections, the ratio of web area to total area is larger than for hot-rolled sections. The inelastic reserve capacity of cold-formed sections may therefore be greater than that for doubly symmetric hot-rolled sections. Consequently an allowance for inelastic reserve capacity may produce substantial gains in capacity for cold-formed sections even if the maximum strain in the section is limited in keeping with the reduced ductility of cold-formed steel sections.

Tests on cold-formed hat sections were reported in Ref. 5.19 and are shown in Figure 5.18a where the ratio (C_y) of the maximum compressive strain to the yield strain has been plotted against the compression flange slenderness for a range of section slenderness values. An approximate lower bound to these tests is shown, which is the basis of Section C3.1.1(b) in the AISI Specification. For sections

155

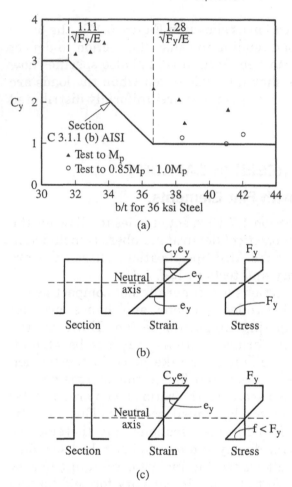

(a)

(b)

(c)

FIGURE 5.18 Stresses and strains for assessing inelastic reserve capacity: (a) compression strain factor for stiffened compression elements; (b) stress distribution for tension and compression flange yielded; (c) stress distribution for tension flange not yielded.

with stiffened compression flanges with slenderness less than $1.11/\sqrt{F_y/E}$, a maximum value of compression flange strain equal to three times the yield strain ($C_y = 3.0$) is allowed. For sections with stiffened compression flanges with slenderness greater than $1.28/\sqrt{F_y/E}$, the maximum compression flange strain is the yield strain ($C_y = 1.0$). For intermediate slenderness values, linear interpolation should be used.

From these values of C_y, the stress and strain distributions in Figures 5.18b and 5.18c can be determined. The stress distributions assume an ideally elastic-plastic material. The ultimate moment (M_u) causing a maximum compression strain of $C_y e_y$ can then be calculated from the stress distributions. Section C3.1.1(b) of the AISI Specification limits the nominal flexural strength (M_n) to $1.25 S_e F_y$. This limit requires that conditions (1)–(5) of Section C3.1.1(b) be satisfied. These limits ensure that other modes of failure such as torsional-flexural buckling or web failure do not occur before the ultimate moment is reached. For sections with unstiffened compression flanges and multiple stiffened segments, Section C3.1.1(b) of the AISI Specification requires that $C_y = 1$.

5.5.2 Cylindrical Tubular Members

Section C6.1 of the AISI Specification gives design rules for cylindrical tubular members in bending and includes the inelastic reserve capacity for compact sections for which $M_n = 1.25 F_y S_f$.

The effective section modulus (S_e) nondimensionalized with respect to the full elastic section modulus (S_f) has been plotted in Figure 5.19 versus the section slenderness (λ_s) based on the outside diameter-to-thickness ratio (D/t) and the yield stress (F_y). A test database provided by Sherman (Ref. 5.20) for constant moment and cantilever tests on electric resistance welds (ERW) and fabricated tubes has also been plotted on the figure. The most significant feature of the AISI Specification is that the full-section plastic moment capacity can be used for compact section tubes.

157

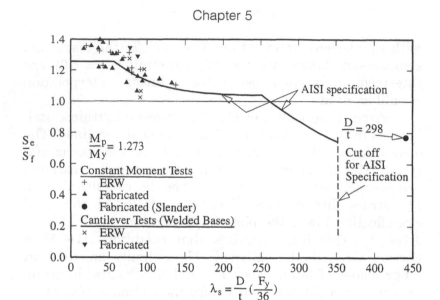

$$\lambda_s = \frac{D}{t}\left(\frac{F_y}{36}\right)$$

FIGURE 5.19   Circular hollow sections (bending).

5.6 EXAMPLE: SIMPLY SUPPORTED C-SECTION BEAM

Problem

Determine the design load on the C-section beam in Example 4.6.3 simply supported over a 25 ft. span with one brace at the center and loaded on the tension flange as shown in Figure 5.20. Use the lateral buckling strength (Section 3.1.2). The following section, equation, and table numbers refer to those in the AISI Specification.

$$F_y = 55 \text{ ksi} \qquad E = 29500 \text{ ksi} \qquad G = 11300 \text{ ksi}$$
$$L = 300 \text{ in.} \qquad H = 8 \text{ in.} \qquad B = 2.75 \text{ in.}$$
$$R = 0.1875 \text{ in.} \qquad D = 0.625 \text{ in.} \qquad t = 0.060 \text{ in.}$$

Flexural Members

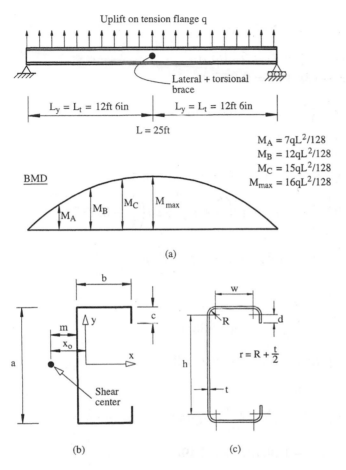

$$M_A = 7qL^2/128$$
$$M_B = 12qL^2/128$$
$$M_C = 15qL^2/128$$
$$M_{max} = 16qL^2/128$$

(a)

(b)   (c)

FIGURE 5.20   C-section beam loading and dimensions: (a) loading and bending moment distribution; (b) centerline dimensions; (c) flat dimensions.

Solution

I.   Area, Major, and Minor Axis Second Moments of Area and Torsion Constant of Full Section accounting for Rounded Corners (see Figure 5.20c)

$$r = R + \frac{t}{2} = 0.2175 \text{ in.}$$

$$h = H - (2r + t) = 7.505 \text{ in.}$$

$$w = B - (2r + t) = 2.255 \text{ in.}$$

$$d = D - \left(r + \frac{t}{2}\right) = 0.378 \text{ in.}$$

$$u = \frac{\pi r}{2} = 0.342 \text{ in.}$$

$$A = t(h + 2w + 4u + 2d) = 0.848 \text{ in.}^2$$

$$I_x = 2t\left[0.0417h^3 + w\left(\frac{h}{2} + r\right)^2 + 2u\left(\frac{h}{2} + 0.637r\right)^2\right.$$

$$\left. + 2(0.149r^3) + 0.0833d^3 + \frac{d}{4}(h - d)^2\right]$$

$$= 8.198 \text{ in.}^4$$

$$\bar{x} = \frac{2t}{A}\left[w\left(\frac{w}{2} + r\right) + u(0.363r)\right.$$

$$\left. + u(w + 1.637r) + d(w + 2r)\right]$$

$$= 0.703 \text{ in.}$$

$$I_y = 2t\left[w\left(\frac{w}{2} + r\right)^2 + \frac{w^3}{12} + 0.356r^3 + d(w + 2r)^2\right.$$

$$\left. + u(w + 1.637r)^2 + 0.149r^3\right]$$

$$- A\bar{x}^2$$

$$= 0.793 \text{ in.}^4$$

$$J = \frac{t^3}{3}(h + 2w + 2d + 4u) = 0.00102 \text{ in.}^4$$

II. Warping Constant and Shear Center Position for Section with Square Corners (see Figure 5.20b)

$$a = H - t = 7.94 \text{ in.}$$

$$b = B - t = 2.69 \text{ in.}$$

$$c = D - \frac{t}{2} = 0.595 \text{ in.}$$

$$m = \frac{b(3a^2b + c(6a^2 - 8c^2))}{a^3 + 6a^2b + c(8c^2 - 12ac + 6a^2)}$$

$$= 1.151 \text{ in.}$$

$$C_w = \frac{a^2b^2t}{12}$$

$$\times \left\{ \frac{\begin{array}{cccc} 2a^3b & +3a^2b^2 + 48c^4 & +112bc^3 & +8ac^3 \\ +48abc^2 & +12a^2c^2 & +12a^2bc & +6a^3c \end{array}}{6a^2b + (a + 2c)^3 - 24ac^2} \right\}$$

$$= 10.362 \text{ in.}^6$$

$$x_o = -(\bar{x} + m) = -1.854 \text{ in.}$$

III. Design Load on Braced Beam
   Section C3.1.2 Lateral buckling strength
   A. Effective Lengths for Central Brace

$$L_y = \frac{L}{2} = 150 \text{ in.} \qquad L_t = \frac{L}{2} = 150 \text{ in.}$$

   B. Buckling Stresses

$$r_y = \sqrt{\frac{I_y}{A}} = 0.967 \text{ in.}$$

$$\sigma_{ey} = \frac{\pi^2 E}{(L_y/r_y)^2} = 12.10 \text{ ksi} \qquad \text{(Eq. C3.1.2-9)}$$

$$r_o = \sqrt{\frac{I_x + I_y}{A} + x_o^2} = 3.747 \text{ in.} \qquad \text{(Eq. C3.1.2-13)}$$

$$\sigma_t = \frac{GJ}{Ar_0^2}\left(1 + \frac{\pi^2 E C_w}{GJL_t^2}\right) = 12.23 \text{ ksi} \text{(Eq. C3.1.2-10)}$$

C. C_b Factor for Uniformly Distributed Load (see Figure 5.20a)

$$C_b = \frac{12.5 M_{\max}}{2.5 M_{\max} + 3M_A + 4M_B + 3M_C} = 1.299$$

(Eq. C3.1.2-11)

$$M_e = C_b A r_o \sqrt{\sigma_{ey}\sigma_t} = 50.202 \text{ kip-in.} \quad \text{(Eq. C3.1.2-6)}$$

C. Critical Moment (M_c)

$$S_f = \frac{I_x}{H/2} = 2.049 \text{ in.}^3$$

$$M_y = S_f F_y = 112.72 \text{ kip-in.} \qquad \text{(Eq. C3.1.2-5)}$$

$$M_c = M_e = 50.20 \text{ kip-in.} \qquad \text{since } M_e < 0.56\,M_y$$

(Eq. C3.1.2-4)

$$F_c = \frac{M_c}{S_f} = 24.50 \text{ ksi}$$

D. Nominal Strength of Laterally Unbraced Segment (M_n)

S_c is the effective section modulus at the stress $F_c = M_c/S_f$. Use calculation method in Example 4.6.3 with $f = 24.495$ ksi.

$$S_c = 1.949 \text{ in.}^3$$

$$M_n = S_c F_c = 47.741 \text{ kip-in.}$$

LRFD $\qquad \phi_b = 0.90$

$$q_D = \frac{8\phi_b M_n}{L^2} = 3.819 \text{ lb/in.}$$

ASD $\qquad \Omega_b = 1.67$

$$q_D = \frac{8 M_n}{\Omega_b L^2} = 2.541 \text{ lb/in.}$$

REFERENCES

5.1 Galambos, T. V., Structural Members and Frames, Prentice-Hall, Englewood Cliffs, New Jersey, 1968.

5.2 Trahair, N. S., Flexural-Torsional Buckling, E. & F. N. Spon, 1993.

5.3 Ings, N. L. and Trahair, N. S., Lateral buckling of restrained roof purlins, Thin-Walled Structures, Vol. 2, No. 4, 1984, pp. 285–306.

5.4 Hancock, G. J. and Trahair, N. S., Finite element analysis of the lateral buckling of continuously restrained beam-columns, Civil Engineering Transactions, Institution of Engineers, Australia, Vol. CE20, No. 2, 1978.

5.5 Hancock, G. J. and Trahair, N. S., Lateral buckling of roof purlins with diaphragm restraints, Civil Engineering Transactions, Institution of Engineers, Australia, Vol. CE21, No. 1, 1979.

5.6 Users Manual for Program PRFELB, Elastic Flexural-Torsional Buckling Analysis, Version 3.0, Center for Advanced Structural Engineering, Department of Civil Engineering, University of Sydney, 1997.

5.7 Kirby, P. A. and Nethercot, D. A., Design for Structural Stability, Wiley, New York, 1979.

5.8 Galambos, T. V., Inelastic buckling of beams, Journal of the Structural Division, ASCE, 89(ST5), October 1963.

5.9 Peköz, T. B. and Sumer, O., Design Provisions for Cold-Formed Steel Columns and Beam-Columns, Final Report, Cornell University, submitted to American Iron and Steel Institute, September 1992.

5.10 Kian, T. and Peköz, T. B., Evaluation of Industry-Type Bracing Details for Wall Stud Assemblies, Final Report, Cornell University, submitted to American Iron and Steel Institute, January 1994.

5.11 Trahair, N. S., Lateral buckling strengths of unsheeted cold-formed beams, Engineering Structures, 16(5), 1994, pp. 324–331.

5.12 Baigent, A. H. and Hancock, G. J., Structural analysis of assemblages of thin-walled members, Engineering Structures, Vol. 4, No. 3, 1982.

5.13 Hancock, G. J., Portal frames composed of cold-formed channel and Z-sections, in Steel Framed Structures—Stability and Strength, (R. Narayanan, ed.), Elsevier Applied Science Publishers, London, 1985, Chap. 8.

5.14 Vlasov, V. Z., Thin-Walled Elastic Beams, Moscow (English translation, Israel Progam for Scientific Translation, Jerusalem), 1961.

5.15 Peköz, T. and P. Soroushian, Behavior of C- and Z-Purlins under Wind Uplift, Sixth International Specialty Conference on Cold-Formed Steel Structures, St Louis, MO, Nov. 1982.

5.16 Rousch, C. J. and Hancock, G. J., Comparison of a non-linear Purlin Model with Tests, Proceedings 12th International Specialty Conference on Cold-Formed Steel Structrures, St. Louis, MO, 1994, pp. 121–149.

5.17 Rousch, C. J. and Hancock, G. J., Determination of Purlin R-Factors using a Non-linear Analysis, Proceedings 13th International Speciality Conference on Cold-Formed Steel Structures, St. Louis, MO, 1996, pp. 177–206.

5.18 Rousch, C. J. and Hancock, G. J., Comparison of tests of bridged and unbridged purlins with a non-linear analysis model, J. Constr. Steel Res., Vol. 41, No. 2/3, 1997, pp. 197–220.

5.19 Reck, P. H., Peköz, T., and Winter, G., Inelastic strength of cold-formed steel beams, Journal of the Structural Division, ASCE, Vol. 101, No. ST11, Nov. 1975.

5.20 Sherman, D. R., Inelastic flexural buckling of cylinders, in Steel Structures—Recent Research Advances and Their Applications to Design (M.N. Pavlovic, ed.), London, Elsevier Applied Science Publishers, 1986, pp. 339–357.

6

Webs

6.1 GENERAL

The design of webs for strength is usually governed by

(a) The web subjected to shear force and undergoing shear buckling, or yielding in shear or a combination of the two, or

(b) The web subjected to in-plane bending and undergoing local buckling in the compression zone, or

(c) The web subjected to a concentrated transverse force (edge load) at a load or reaction point and undergoing web crippling

In all three cases, the buckling modes may interact with each other to produce a lower strength if the shear and bending or the bending moment and concentrated load occur simultaneously.

In the AISI Specification, the design of webs in shear is governed by Section C3.2. The interaction between shear and bending is covered in Section C3.3. The design of webs for web crippling under point loads and reactions is governed by Section C3.4, and combined bending and web crippling is prescribed in Section C3.5. The design of transverse stiffeners at bearing points and as intermediate stiffeners is set out in Section B6.

6.2 WEBS IN SHEAR

6.2.1 Shear Buckling

The mode of buckling in shear is shown in Figure 6.1a. This mode of buckling applies for a long plate which is simply supported at the top and bottom edges. Unlike local buckling in compression, the nodal lines are not perpendicular to the direction of loading and parallel with the loaded edges of the plate. Consequently, the basic mode of buckling is altered if the plate is of finite length. The buckling stress for shear buckling is given by Eq. (4.1), with the value of the local buckling coefficient (k) selected to allow for the shear loading and boundary conditions. The theoretical values of k as a function of the aspect ratio (a/b) of the plate are shown in Figure 6.1b. As the plate is shortened, the number of local buckles is reduced and the value of k is increased from 5.34 for a very long plate to 9.34 for a square plate. An approximate formula for k as a function of a/b is shown by the solid line in Figure 6.1b and given by

$$k = 5.34 + \frac{4}{(a/b)^2} \qquad (6.1)$$

Equation (6.1) is a parabolic approximation to the garland curve in Figure 6.1b.

For plates with intermediate transverse stiffeners along the web, the effect of the stiffeners is to constrain the shear buckling mode between stiffeners to be the same as a plate of length equal to the distance between the stiffeners. Consequently, the value of a in Eq. (6.1) is

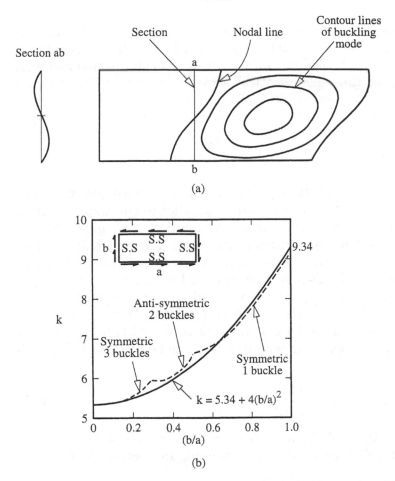

FIGURE 6.1 Simply supported plate in shear: (a) buckled mode in shear for long narrow plate; (b) buckling coefficient (*k*) versus aspect ratio (*a/b*).

taken as the stiffener spacing and an increased value of the local buckling coefficient over that for a long unstiffened plate can be used for webs with transverse stiffeners.

As explained in Ref. 6.1, plates in shear are unlikely to have a substantial postbuckling reserve as for plates in compression unless the top and bottom edges are held straight. Since cold-formed members usually only have thin flanges, it is unlikely that the postbuckling strength in shear can be mobilized in the web of a cold-formed member. Hence, the elastic critical stress is taken as the limiting stress in Section C3.2 of the AISI Specification. Section C3.2 uses Eq. (4.1) to give a nominal shear strength (V_n) of

$$V_n = \frac{0.905Ek_v}{(h/t)^2} ht = \frac{0.905Ek_v t^3}{h} \tag{6.2}$$

where h is web depth, which is taken as the depth of the flat portion of the web measured parallel to the web, and k_v is taken as the parabolic approximation to k given by Eq. (6.1). When the stiffener spacing is greater than the web depth such that $a/h > 1.0$, then b in Eq. (6.1) is replaced by h. When the stiffener spacing is less than the web depth such that $a/h \leq 1.0$, then the plate is taken as a plate of length h with a flat width ratio of a/t. Hence, w/t in Eq. (4.1) is replaced by a/t, and a/b in Eq. (6.1) is replaced by h/a. The resulting equation for k_v to be used with Eq. (6.2) is

$$k_v = 4.0 + \frac{5.34}{(a/h)^2} \tag{6.3}$$

For plates without stiffeners, k_v is simply taken as 5.34. These formulae for k_v are specified in Section C3.2.

6.2.2 Shear Yielding

A stocky web (small h/t) subject to shear will yield in shear at an average stress of approximately $F_y/\sqrt{3}$, as given by the von Mises' yield criterion (Ref. 6.2). The nominal shear capacity for yield (V_n) is therefore

$$V_n = 0.60F_y ht \tag{6.4}$$

The value of 0.60, which is higher than $1/\sqrt{3}$, is consistent with a reduced factor of safety normally assumed for shear yielding in allowable stress design standards such as the

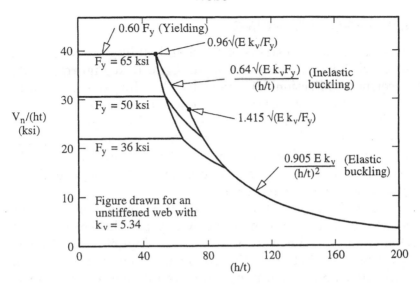

FIGURE 6.2 Nominal shear strength (V_n) of a web expressed as average shear stress.

AISC Specification (Ref. 1.1). Figure 6.2 shows the failure stresses for a web in shear, including yield as given by Eq. (6.4) and buckling as given by Eq. (6.2). In the region where shear and buckling interact, the failure stress is given by the geometric mean of the buckling stress and 0.8× the yield stress in shear (Ref. 6.3). The resulting equation for the nominal shear strength (V_n) is given by Eq. (6.5a) and is shown in Figure 6.2 for a range of yield stresses:

$$V_n = 0.64t^2\sqrt{Ek_vF_y} \qquad (6.5a)$$

This equation applies in the range of web slenderness:

$$0.96\sqrt{\frac{Ek_v}{F_y}} \leq \frac{h}{t} \leq 1.415\sqrt{\frac{Ek_v}{F_y}} \qquad (6.5b)$$

The resistance factors (ϕ_v) and safety factors (Ω_v) are 1.0 and 1.5, respectively, for Eq. (6.4), and 0.90 and 1.67, respectively, for Eqs. (6.2) and (6.5).

6.3 WEBS IN BENDING

The elastic critical stress of a web in bending is given in Eq. (4.1) with $k = 23.9$, as shown as Case 5 in Figure 4.1. As described in Section 4.4, the web in bending has a substantial postbuckling reserve. Consequently, the design procedure for webs in bending should take account of the postbuckling reserve of strength. Two basic design methods have been described in Ref. 6.4. Both methods result in approximately the same design strength but require different methods of computation.

The first procedure extends the concept of effective width to include an effective width for the compression component of the web as shown in Figure 6.3a. Consequently the effective section involves removal of portions of both the compression flange and web if each is sufficiently slender. The effective width formulae for b_1 and b_2 in the web, as shown in Figure 6.3a, have been proposed in Ref. 6.4 and confirmed by tests. They are described in Section 4.4 and demonstrated in Example 4.6.3. The disadvantage of the method is the complexity of the calculation, including the determination of the centroid of the effective section. The 1996 AISI Specification uses this design approach.

An alternative procedure has also been proposed in Ref. 6.4 where only the ineffective component of the compression flange has been removed and the full web depth is considered as shown in Figure 6.3b. In this case, a limiting stress (F_{bw}) in the web in bending is specified. The method is simpler to use than the effective web approach since it is not iterative. It was selected for the 1980 edition of the AISI Specification as well as the British Standard BS 5950 (Part 5) (Ref. 1.5). The resulting formulae for the maximum permissible stress in a web in bending are Eqs. (6.6) and (6.7).

For sections with stiffened compression flanges:

$$F_{bw} = \left[1.21 - 0.00034\left(\frac{h}{t}\right)\sqrt{F_y}\right]F_y \qquad (6.6)$$

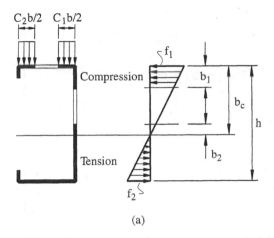

(a)

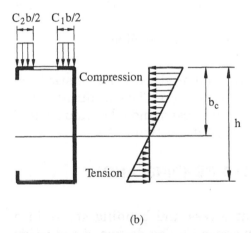

(b)

FIGURE 6.3 Effective width models for thin-walled sections in bending: (a) effective compression flange and web; (b) effective compression flange and full web.

For sections with unstiffened compression flanges:

$$F_{bw} = \left[1.26 - 0.00051\left(\frac{h}{t}\right)\sqrt{F_y}\right]F_y \qquad (6.7)$$

Both formulae have been confirmed by the testing reported in Ref. 6.4. The resulting design curves are plotted in

173

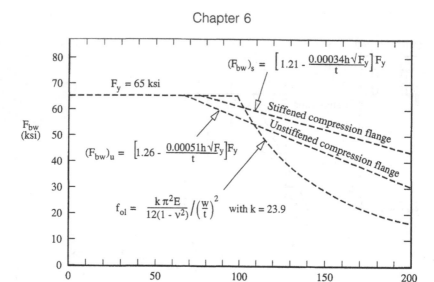

FIGURE 6.4   Limiting stress on a web in bending.

Figure 6.4, where they are compared with the elastic local buckling stress for a web in bending. The limiting stress F_{bw} has been used in the purlin design model of Rousch and Hancock (Ref. 5.18), described in Section 5.3.2.

6.4 WEBS IN COMBINED BENDING AND SHEAR

The combination of shear stress and bending stress in a web further reduces web capacity. The degree of reduction depends on whether the web is stiffened. The interaction formulae for combined bending and shear are given in Section C3.3 of the AISI Specification.

For an unstiffened web the interaction equation is a circular formula as shown in Figure 6.5a. This interaction formula is based upon an approximation to the theoretical interaction of local buckling resulting from shear and bending as derived by Timoshenko and Gere (Ref. 3.6). As applied in the AISI Specification, the interaction is based on the nominal flexural strength (M_{nxo}), which includes

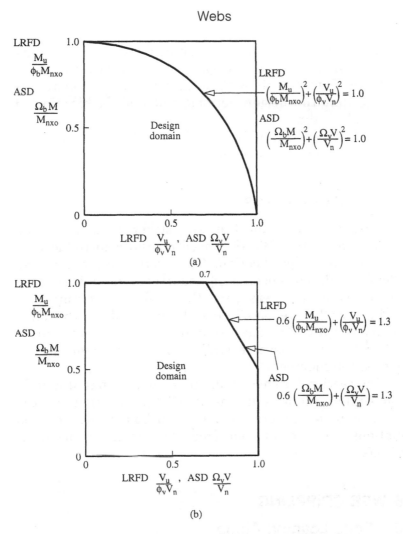

Webs

(a)

$$\left(\frac{M_u}{\phi_b M_{nxo}}\right)^2 + \left(\frac{V_u}{\phi_v V_n}\right)^2 = 1.0$$

LRFD

$$\left(\frac{\Omega_b M}{M_{nxo}}\right)^2 + \left(\frac{\Omega_v V}{V_n}\right)^2 = 1.0$$

ASD

LRFD $\dfrac{V_u}{\phi_v V_n}$, ASD $\dfrac{\Omega_v V}{V_n}$

(b)

LRFD

$$0.6\left(\frac{M_u}{\phi_b M_{nxo}}\right) + \left(\frac{V_u}{\phi_v V_n}\right) = 1.3$$

ASD

$$0.6\left(\frac{\Omega_b M}{M_{nxo}}\right) + \left(\frac{\Omega_v V}{V_n}\right) = 1.3$$

LRFD $\dfrac{V_u}{\phi_v V_n}$, ASD $\dfrac{\Omega_v V}{V_n}$

FIGURE 6.5 Combined bending and shear in webs: (a) unstiffened webs; (b) webs with transverse stiffeners.

post–local buckling in bending. Hence, the justification for the use of the circular formula is actually empirical as confirmed by the testing reported in Ref. 6.5.

For a stiffened web, the interaction between shear and bending is not as severe, probably as a result of a greater postbuckling capacity in the combined shear and bending buckling mode. Consequently, a linear relationship, as shown in Figure 6.5b with a larger design domain, is used to limit the design actions under a combination of shear and bending.

In Figures 6.5a and b, the LRFD and ASD formulae in the AISI Specification are shown.

6.5 WEB STIFFENERS

The design sections for the web stiffener requirements specified in Section B6 are based on those introduced in the 1980 AISI Specification. The section for transverse stiffeners (B6.1) has been designed to prevent end crushing of transverse stiffeners [Eq. (B6.1-1)] and column-type buckling of the web stiffeners. It is based on the tests described in Ref. 6.6. The resistance factor (ϕ_c) and safety factor (Ω_c) for transverse stiffeners are the same as for compression members.

The section for shear stiffeners (B6.2) is based mainly upon similar sections in the AISC Specification for the design of plate girders (Ref. 6.7), although the detailed equations were confirmed from the tests reported in Ref. 6.6.

6.6 WEB CRIPPLING

6.6.1 Edge Loading Alone

In the design of cold-formed sections, it is not always possible to provide load-bearing stiffeners at points of concentrated edge loading. Consequently, a set of rules is given in Section C3.4 of the AISI Specification for the design against web crippling under concentrated edge load in the manner shown in Figure 6.6. The equations in the AISI Specification have been empirically based on tests. The nominal web crippling strength (P_n) has been found to

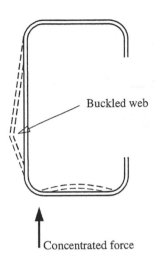

FIGURE 6.6 Web crippling of an open section.

be a function of the following parameters as shown in Figure 6.7. These are:

(a) The nature of the restraint to web rotation provided by the flange and adjacent webs as shown in Figure 6.7a. The back-to-back channel beam has a higher restraint to web rotation at the top and bottom, and hence a higher load capacity than a single channel. Similarly, the channel with a stiffened, or partially stiffened, compression flange has a higher restraint to web rotation than the channel with an unstiffened flange, and hence a higher web-bearing capacity.

(b) The length of the bearing (N) shown in Figure 6.7b and its proximity to the end of the section (defined by c). In addition the proximity of other opposed loads, defined by e in Figure 6.7b, is also important. A limiting value of c/h of 1.5, where h is the web flat width, is used to distinguish between end loads and interior loads. Similarly a limiting value of $e/h = 1.5$ is used to distinguish between opposed loads and nonopposed loads.

177

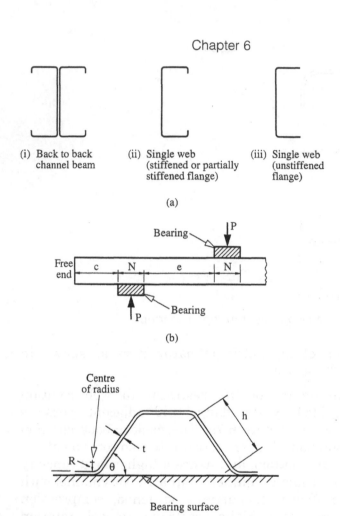

(i) Back to back
    channel beam

(ii) Single web
     (stiffened or partially
     stiffened flange)

(iii) Single web
      (unstiffened
      flange)

(a)

Bearing

P

Free
end

c   N   e   N

Bearing

P

(b)

Centre
of radius

h

t

R

θ

Bearing surface

(c)

FIGURE 6.7 Factors affecting web crippling strength: (a) restraint against web rotation; (b) bearing length and position; (c) section geometry.

(c)   The web thickness (t), the web slenderness (h/t), the web inclination (θ), and the inside bend radius (R), as shown in Figure 6.7c, are the relevant section parameters defining the section geometry. The following limits to the geometric parameters apply to Section 3.4.

Web slenderness: $h/t \leq 200$
Web inclination: $90° \geq \theta \geq 45°$
Bend radius to thickness: $R/t \leq 6$ for beams and
  ≤ 7 for sheathing
Bearing length to thickness: $N/t \leq 210$
Bearing length to web depth: $N/h \leq 3.5$

Transverse stiffeners must be used for webs whose slenderness exceeds 200. The factors ϕ_w and Ω_w for single unreinforced webs are 0.75 and 1.85, respectively, and for I-sections are 0.80 and 2.0, respectively. For two nested Z-sections, when evaluating the web crippling strength for interior one flange loading ($c/h > 1.5$, $e/h > 1.5$), ϕ_w and Ω_w are 0.85 and 1.80, respectively (Ref. 6.8).

6.6.2 Combined Bending and Web Crippling

The combination of bending moment and concentrated force frequently occurs in beams at points such as interior supports and points of concentrated load within the span. When bending moment and concentrated load occurs simultaneously at points without transverse stiffeners, the two actions interact to produce a reduced load capacity.

A large number of tests have been performed, mainly at the University of Missouri-Rolla (UMR) and at Cornell University, to determine the extent of this interaction. The test results (Refs. 6.9, 6.10) are summarized in Figure 6.8 for sections with single webs and in Figure 6.9 for I-beams formed from back-to-back channels. In the latter case, the test results only apply for beams which have webs with a slenderness (h/t) greater than $2.33/\sqrt{F_y/E}$, or with flanges which are not fully effective, or both. For I-beams with stockier flanges and webs, no significant interaction between bending and web crippling occurs, as shown in Ref. 6.8.

The test results in Figures 6.9 and 6.10 have been nondimensionalized with respect to the theoretical ultimate moments and concentrated loads, which have been computed by using Section C3.1.1 and Section C3.4. For the

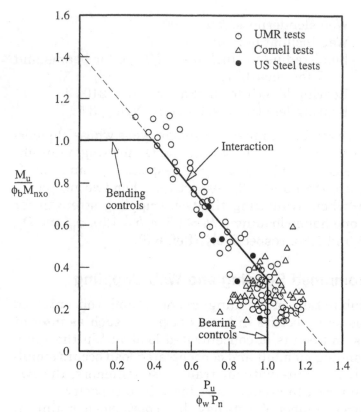

FIGURE 6.8 Interaction of bending and bearing for single webs (LRFD formulation).

purpose of the design equations, the relevant resistance factors (ϕ_b and ϕ_w) are included in the nondimensionalization in Figures 6.8 and 6.9. The relevant design formulae follow:

(a) For shapes having single unreinforced webs:

$$\text{LRFD} \qquad 1.07\left(\frac{P_u}{\phi_w P_n}\right) + \frac{M_u}{\phi_b M_{nxo}} \leq 1.42 \qquad (6.8a)$$

$$\text{ASD} \qquad 1.2\left(\frac{\Omega_w P}{P_n}\right) + \frac{\Omega_b M}{M_{nxo}} \leq 1.5 \qquad (6.8b)$$

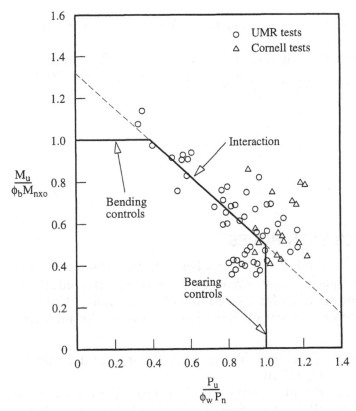

Webs

FIGURE 6.9 Interaction of bending and bearing for webs of I-beams (LRFD formulation).

(b) For back-to-back channel beams and beams with restraint against web rotation:

$$\text{LRFD} \qquad 0.82\frac{P_u}{\phi_w P_n} + \frac{M_u}{\phi_b M_{nxo}} \le 1.32 \qquad (6.9a)$$

$$\text{ASD} \qquad 1.1\left(\frac{\Omega_w P}{P_n}\right) + \frac{\Omega_b M}{M_{nxo}} \le 1.5 \qquad (6.9b)$$

Recent research at the University of Missouri-Rolla and the University of Wisconsin-Milwaukee (Ref. 6.8) has led to specific formulae for the case of two nested Z-shapes. The

181

web crippling and bending behavior are enhanced by the interaction of the nested webs.

(c) For the support point of two nested Z-shapes:

$$\text{LRFD} \qquad \frac{M_u}{M_{no}} + \frac{P_u}{P_n} \leq 1.68\phi \qquad\qquad (6.10a)$$

$$\text{ASD} \qquad \frac{M}{M_{no}} + \frac{P}{P_n} \leq \frac{1.67}{\Omega} \qquad\qquad (6.10b)$$

where $\phi = 0.9$ and $\Omega = 1.67$.

6.7 WEBS WITH HOLES

In the July 1999, Supplement No. 1 to the 1996 edition of the AISI Specification, significant revisions and additions were made to allow for C-section webs with holes. Changes to the 1996 Specification have been made in the following sections:

B2.4   C-Section Webs with Holes under Stress Gradient

C3.2.2   Shear Strength of C-Section Webs with Holes

C3.4.2   Web Crippling Strength of C-Section Webs with Holes

In all cases, the following limits apply:

1. $d_o/h < 0.7$.
2. $h/t \leq 200$.
3. Holes centered at mid-depth of the web.
4. Clear distance between holes ≥ 18 in. (457 mm).
5. Noncircular holes, corner radii $\geq 2t$.
6. Noncircular holes, $d_o \leq 2.5$ in. (64 mm) and $b \leq 4.5$ in. (114 mm), where $d_o =$ depth of web hole as shown in Figure 6.10 and $b =$ length of web hole as shown in Figure 6.10.
7. Circular hole diameters ≤ 6 in. (152 mm).
8. $d_o > 9/16$ in. (14 mm).

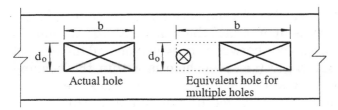

FIGURE 6.10 Holes in webs.

When multiple holes occur, an equivalent hole can be defined bounding the multiple holes as shown in Figure 6.10.

In the case of Section B2.4, when $d_o/h < 0.38$, the effective widths (b_1 and b_2) in Figure 6.3a shall be determined as discussed in Section 4.4 of this book and Section B2.3(a) of the AISI Specification by assuming no hole exists in the web. When $d_o/h \geq 0.38$, the effective width shall be determined by Section B3.1(a) of the AISI Specification assuming the compression portion of the web consists of an unstiffened element adjacent to the hole with $f = f_1$ as shown in Figure B2.3-1 of the AISI Specification and Figure 6.3a of this book. These conclusions were based on the research of Shan et al. (Ref. 6.11). For rectangular holes which exceed the limits for noncircular holes (limit 6 above), an equivalent virtual circular hole which circumscribes the rectangular hole may be used.

In the case of Section C3.2.2, the nominal shear strength (V_n) described in Section 6.2 of this book and Section C3.2.1 of the AISI Specification is reduced by a factor q_s which depends upon the h and t of the web and d_o of the hole, as shown in Figure 6.10. These reduction factors were based on research by Shan et al. (Ref. 6.11), Schuster et al. (Ref. 6.12), and Eiler et al. (Ref. 6.13).

In the case of Section C3.4.2, the nominal web crippling strength (P_n) described in Section 6.6 of this book and

Section C3.4 of the AISI Specification is reduced by a factor R_c which depends upon h and the hole depth (d_o) as well as the nearest distance x between the web hole and the edge of bearing. When $x \leq 0$ (i.e., web hole is within the bearing length), a bearing stiffener must be used. These reduction factors are based on research by Langan et al. (Ref. 6.14), Uphoff (Ref. 6.15), and Deshmukh (Ref. 6.16).

6.8 EXAMPLE: COMBINED BENDING WITH WEB CRIPPLING OF HAT SECTION

Problem

Determine the design web crippling strength of the hat section in Example 4.6.1 for a bearing length of $N = 2$ in. at an interior loading point. Determine also the design flexural strength at the loading point when the load is half of the design web cripping strength computed. The following section and equation numbers refer to the AISI Specification.

Solution

A.   Web Crippling over an Interior Support
     Section C3.4 Web crippling strength

$$\frac{R}{t} = \frac{0.125}{0.060} = 2.08 < 6$$

(Table C3.4-1)

   Shapes having single webs and stiffened flanges: single load or reaction point $c > 1.5h$

$$P_n = t^2 k C_1 C_2 C_9 C_\theta \left[538 - 0.74\frac{h}{t} \right]\left[1 + 0.007\frac{N}{t} \right]$$

(Eq. C3.4-4)

where

$$k = \frac{894F_y}{E} = 1.515$$

$$C_1 = 1.22 - 0.22k = 0.887$$

$$C_2 = 1.06 - 0.06\frac{R}{t} = 0.935$$

$C_9 = 1.0$ for U.S. customary units, kips and in.

$$\frac{N}{t} = \frac{2.0}{0.060} = 33.33$$

$$\frac{h}{t} = \frac{4.0 - 2(0.125 + 0.060)}{0.060} = 62.5$$

$$C_\theta = 0.7 + 0.3\left(\frac{\theta}{90}\right)^2$$

$$= 1.0 \qquad \text{for } \theta = 90°$$

P_n per web $= 2.74$ kips

P_n for two webs $= 5.48$ kips

LRFD $\phi_w P_n$ for two webs is $0.75 \times 5.48 = 4.11$ kips

ASD $\dfrac{P_n}{\Omega_w}$ for two webs is $\dfrac{5.48}{1.85} = 2.96$ kips

B. Combined Bending and Web Crippling
Section C3.5.2 LRFD method
Shapes having single unreinforced webs

$$1.07\left(\frac{P_u}{\phi_w P_n}\right) + \left(\frac{M_u}{\phi_b M_{nxo}}\right) \le 1.42 \qquad \text{(Eq. C3.5.2-1)}$$

For

$$\frac{P_u}{\phi_w P_n} = 0.5, \qquad \frac{M_u}{\phi_b M_{nxo}} = 1.42 - 1.07\,(0.5) = 0.885$$

Hence,

$$M_u = 0.885\phi_b M_{nxo} = 0.885 \times 49.834 \text{ kip-in. (Example 4.6.1)}$$
$$= 44.10 \text{ kip-in.}$$

The bending moment is reduced by 11.5% as a result of a vertical load equal to half of the interior web crippling strength.

Section C3.5.1 ASD method
Shapes having single unreinforced webs:

$$1.2\left(\frac{\Omega_w P}{P_n}\right) + \left(\frac{\Omega_b M}{M_{nxo}}\right) \le 1.5 \qquad \text{(Eq. C.3.5.1-1)}$$

For

$$\left(\frac{\Omega_w P}{P_n}\right) = 0.5, \frac{\Omega_b M}{M_{nxo}} = 1.5 - 1.2(0.5) = 0.90$$

Hence,

$$M = 0.90\frac{M_{nxo}}{\Omega_b} = 0.90 \times 34.411 \qquad \text{(Example 4.6.1)}$$
$$= 30.97 \text{ kip-in.}$$

The bending moment is reduced by 10% as a result of a vertical load equal to half of the interior web crippling strength.

6.9 REFERENCES

6.1 Stein, M., Analytical results for post-buckling behaviour of plates in compression and shear, in Aspects of the Analysis of Plate Structures (Dawe, D. J., Horsington, R. W., Kamtekar, A. G., and Little, G. H., eds., Oxford University Press, 1985, Chap. 12.

6.2 Popov, E. P., Introduction to Mechanics of Solids, Prentice-Hall, Englewood Cliffs, New Jersey, 1968.

6.3 Basler, K., Strength of plate girders in shear, Journal of the Structural Division, ASCE, Vol. 87, No. ST7, Oct. 1961, pp. 151–180.

6.4 LaBoube, R. A. and Yu, W. W., Bending strength of webs of cold-formed steel beams, Journal of the Structural Division, ASCE, Vol. 108, No. ST7, July 1982, pp. 1589–1604.

6.5 LaBoube, R. A. and Yu, W. W., Cold-Formed Steel Web Elements under Combined Bending and Shear, Fourth International Specialty Conference on Cold-Formed Steel Structures, St Louis, MO, June 1978.

6.6 Phung, N. and Yu, W. W., Structural Behaviour of Transversely Reinforced Beam Webs, Final Report, Civil Engineering Study 78-5, University of Missouri-Rolla, Rolla, MO, June 1978.

6.7 American Institute of Steel Construction, Specification for the Design, Fabrication and Erection of Structural Steel in Buildings, Nov. 1978, Chicago, IL.

6.8 LaBoube, R. A. Nunnery, J. N., and Hodges, R. E., Web crippling behavior of nested Z-purlins, Engineering Structures, Vol. 16, No. 5, July 1994.

6.9 Hetrakul, N. and Yu, W. W., Structural Behavior of Beam Webs Subjected to Web Crippling and a Combination of Web Crippling and Bending, Final Report, Civil Engineering Study, 78-4, University of Missouri-Rolla, Rolla, MO, June 1978.

6.10 Hetrakul, N. and Yu, W. W., Cold-Formed Steel I-Beams Subjected to Combined Bending and Web Cripping in Thin-Walled Structures, Recent Technical Advances and Trends in Design, Research and Construction (Rhodes, J. and Walker, A. C., eds.), Granada, 1980.

6.11 Shan, M. Y., LaBoube, R. A., and Yu, W. W., Behavior of Web Elements with Openings Subjected to Bending, Shear and the Combination of Bending and Shear, Final Report, Civil Engineering Series 94.2, Cold-Formed Steel Series, Department of Civil Engineering, University of Missouri-Rolla, 1994.

6.12 Schuster, R. M., Rogers, C. A., and Celli, A., Research into Cold-Formed Steel Perforated C-Sections in

Shear, Progress Report No. 1 of Phase I of CSSBI/IRAP Project, Department of Civil Engineering, University of Waterloo, Waterloo, Canada, 1995.

6.13 Eiler, M. R., LaBoube, R. A., and Yu, W. W., Behavior of Web Elements with Openings Subjected to Linearly Varying Shear, Final Report, Civil Engineering Series 97-5, Cold-Formed Steel Series, Department of Civil Engineering, University of Missouri-Rolla, 1997.

6.14 Langan, J. E., LaBoube, R. A., and Yu, W. W., Structural Behavior of Perforated Web Elements of Cold-Formed Steel Flexural Members Subjected to Web Crippling and a Combination of Web Crippling and Bending, Final Report, Civil Engineering Series, 94-3, Cold-Formed Steel Series, Department of Civil Engineering, University of Missouri-Rolla, 1994.

6.15 Uphoff, C. A., Structural Behavior of Circular Holes in Web Elements of Cold-Formed Steel Flexural Members Subjected to Web Crippling for End-One-Flange Loading, thesis presented to the faculty of the University of Missouri-Rolla in partial fulfillment for the degree Master of Science, 1996.

6.16 Deshmukh, S. U., Behavior of Cold-Formed Steel Web Elements with Web Openings Subjected to Web Crippling and a Combination of Bending and Web Crippling for Interior-One-Flange Loading, thesis presented to the faculty of the University of Missouri-Rolla in partial fulfillment for the degree Master of Science, 1996.

7

Compression Members

7.1 GENERAL

The design of cold-formed members in compression is generally more complex than conventional hot-rolled steel members. This is a result of the additional modes of buckling deformation which commonly occur in thin-walled structural members. These are

(a) Local buckling and post–local buckling of stiffened and unstiffened compression elements, as described in Chapter 4

(b) Flexural, torsional, and torsional-flexural modes of buckling of the whole compression member as shown in Figure 1.13b.

The design for (a) determines the nominal stub column axial strength (P_{no}) described in Section 7.3. The design for (b) gives the nominal axial strength (P_n) given in Section 7.4. The nominal axial strength is defined in Section C4, Concentrically Loaded Compression Members, of the AISI Specification. The elastic buckling stress for flexural,

189

torsional, and torsional-flexural buckling of members in compression is discussed in Section 7.2.

Section C4 of the AISI Specification includes rules which enable the designer to account for these additional buckling modes as well as the normal failure modes of flexural buckling and yielding commonly considered for doubly symmetric hot-rolled compression members as described in the AISC Specification (Ref. 1.1).

7.2 ELASTIC MEMBER BUCKLING

The elastic buckling load for flexural, torsional, and torsional-flexural buckling of a thin-walled member of general cross section, as shown in Figure 7.1a, was derived by Timoshenko and is explained in detail in Timoshenko and Gere (Ref. 3.6) and Trahair (Ref. 5.2). For a member of length L subjected to uniform compression and restrained at its ends by simple supports which prevent lateral displacement of the section perpendicular to its longitudinal axis, as well as twisting rotation, the elastic critical load (P_e) is given by solution of the following three simultaneous equations in the displacement amplitudes A_1, A_2 in the x-, y-directions, respectively, and twist angle A_3 as follows:

$$(P_{ey} - P_e)A_1 - P_e y_0 A_3 = 0 \tag{7.1}$$

$$(P_{ex} - P_e)A_2 + P_e x_0 A_3 = 0 \tag{7.2}$$

$$-P_e y_0 A_1 + P_e x_0 A_2 + r_0^2(P_t - P_e)A_3 = 0 \tag{7.3}$$

where

$$P_{ex} = \pi^2 \frac{EI_x}{L^2} \tag{7.4}$$

$$P_{ey} = \pi^2 \frac{EI_y}{L^2} \tag{7.5}$$

$$P_t = \frac{GJ}{r_0^2}\left(1 + \frac{\pi^2 EC_w}{GJL^2}\right) \tag{7.6}$$

$$r_0^2 = \frac{I_x + I_y}{A} + x_0^2 + y_0^2 \tag{7.7}$$

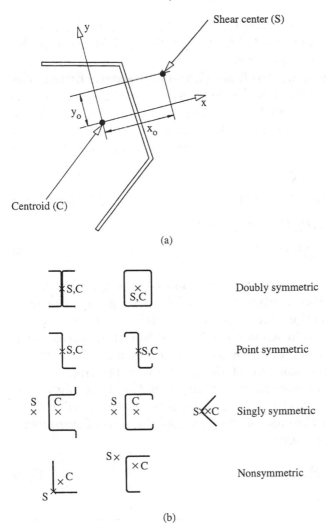

(a)

(b)

FIGURE 7.1 Thin-walled sections: (a) general cross-section; (b) section types.

The terms x_0 and y_0 are the shear center coordinates shown in Figure 7.1a. The modes of buckling for flexure about both principal axes and twist angle are sinusoidal with amplitudes of A_1, A_2, and A_3, respectively. The buckling stresses

191

σ_{ex}, σ_{ey}, and σ_t corresponding to P_{ex}, P_{ey}, and P_t, respectively, are simply derived from Eqs. (7.4), (7.5) and (7.6) by dividing by the gross area.

Two solutions to Eqs. (7.1)–(7.3) exist. Either the column does not buckle and $A_1 = A_2 = A_3 = 0$, or the column buckles and the determinant of the matrix of coefficients of A_1, A_2, and A_3 in Eqs. (7.1)–(7.3) is zero. In this case the resulting cubic equation to be solved for the critical load (P_e) is

$$P_e^3(r_0^2 - x_0^2 - y_0^2) - P_e^2[(P_{ex} + P_{ey} + P_t)r_0^2 - (P_{ey}x_0^2 + P_{ex}y_0^2)]$$
$$+ P_e r_0^2(P_{ex}P_{ey} + P_{ey}P_t + P_tP_{ex})$$
$$- P_{ex}P_{ey}P_t r_0^2 = 0 \qquad (7.8)$$

For general nonsymmetric sections of the type shown in Figure 7.1b, the solution of Eq. (7.8) in its general form is necessary. Section C4.3, Nonsymmetric Sections, of the AISI Specification allows the elastic buckling stress of a nonsymmetric section to be determined by a rational analysis, such as the solution of Eq. (7.8), or by testing.

For singly symmetric sections of the type shown in Figure 7.1b, for which x_0 or y_0 is zero, then simpler solutions exist. In the case of the x-axis as the axis of symmetry, then y_0 is zero; hence

$$(P_e)_1 = P_{ey} \qquad (7.9)$$

$$(P_e)_{2,3} = P_{ext}$$

$$= \frac{(P_{ex} + P_t) \pm \sqrt{(P_{ex} + P_t)^2 - 4P_{ex}P_t[1 - (x_0/r_0)^2]}}{2[1 - (x_0/r_0)^2]}$$

$$(7.10)$$

Equation (7.9) gives the flexural buckling load about the y-axis, and the smaller of the two values computed using Eq. (7.10) gives the torsional-flexural buckling load.

Equation (7.9) for the unlipped channel section in Figure 3.2 produces identical results with the curve through C in Figure 3.3 for column lengths greater than 40 in. where local buckling effects have no influence. Similarly, evaluation of Eq. (7.10) for the channel section in Figure 3.2 produces identical results with the curve through D in Figure 3.3 for column lengths greater than 40 in. In Figures 3.6, 3.7, and 3.9, the curves shown as the dashed line labeled "Timoshenko flexural-torsional buckling formula" were calculated using the lower root of Eq. (7.10).

For doubly symmetric or point symmetric sections where x_0 and y_0 are both zero, the three solutions of Eq. (7.8) are simply

$$(P_e)_1 = P_{ex} \tag{7.11}$$

$$(P_e)_2 = P_{ey} \tag{7.12}$$

$$(P_e)_3 = P_t \tag{7.13}$$

For doubly symmetric sections, closed cross sections, and any other sections that can be shown not to be subject to torsional or torsional-flexural buckling, Section C4.1 of the AISI Specification uses the lesser P_e derived from Eqs. (7.11) and (7.12) divided by the gross area (A) to give the flexural buckling stress (F_e) (Eq. C4.1-1 of the AISI Specification). For singly symmetric sections, where torsional-flexural buckling can also occur, Section C4.2 of the AISI Specification uses P_e derived from Eq. (7.10) divided by A to give F_e. The lesser value of F_e from Section C4.1 and from Section C4.2 must be used for the design of singly symmetric sections. For point symmetric sections, the AISI Specification uses the lesser of P_t derived from Eq. (7.13) divided by the gross area A and F_e from Section C4.1 for flexural buckling alone.

In Eqs. (7.4)–(7.6), L is the unbraced length between the simply supported ends of the column. However, in Eqs. (7.4)–(7.6) L has been replaced by $K_x L_x$, $K_y L_y$, and $K_t L_t$, respectively, in Section C3.1.2 of the AISI Specification. These terms are the effective lengths about the x-, y-, and z-axes, respectively, as defined in Section C3.1.2. The use of different lengths in the computation of the torsional-flexural buckling load by Eqs. (7.8) or (7.10) is not theoretically justifiable. However, it usually produces a conservative estimate of the torsional-flexural buckling load of columns which have different flexural and torsional effective lengths. This procedure has been justified experimentally for the uprights of steel storage rack columns as described in Ref. 7.1.

In Section 6 of the Rack Manufacturers Institute Specification (Ref. 7.2), effective length factors are specified for flexural buckling in the direction perpendicular to the upright frame (typically $K_x = 1.7$ for racking not braced against sidesway), for flexural buckling in the plane of the upright frame (typically $K_y = 1.0$), and for torsional buckling (typically $K_t = 0.8$), provided twisting of the upright is prevented at the brace points.

7.3 STUB COLUMN AXIAL STRENGTH

The strength of short-length columns compressed between rigid end platens (P_{no}) is governed mainly by the yield strength of the material and the slenderness of the plate elements forming the cross section. For sections with stocky plate elements, the strength is simply equal to the squash load given by the product of the full section area (A) and the yield stress (F_y). However, for sections with slender plate elements, the section capacity in compression (often called the stub column strength) is less than the squash load as a result of local buckling.

In Section C4 of the AISI Specification, the stub column strength is defined by

$$P_{no} = A_e F_y \qquad (7.14)$$

where A_e is the effective area of the section at the yield stress F_y computed by summing the effective areas of all

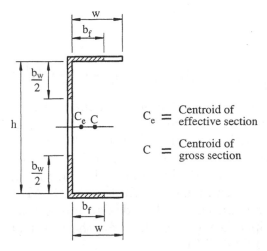

C_e = Centroid of effective section

C = Centroid of gross section

FIGURE 7.2   Effective section for post–locally buckled channel.

the individual elements forming the cross section, as shown for the unlipped channel section in Figure 7.2. The effective areas of the individual elements forming the cross section are computed from the effective width formula [Eq. (4.9)] for each element in isolation, ignoring any interaction which may occur between different elements of the cross section. This approach appears to produce reasonable estimates of the stub column axial strength of thin-walled sections.

7.4 LONG COLUMN AXIAL STRENGTH

The column design curve in the AISI Specification is based upon the American Institute of Steel Construction LRFD Specification curve (Ref. 1.1). The formula for the critical stress (F_n) on a column is given in Section C4 of the AISI Specification as

$$F_n = 0.658^{\lambda_c^2} F_y \qquad \text{for } \lambda_c \leqslant 1.5 \tag{7.15}$$

$$F_n = \frac{0.877}{\lambda_c^2} F_y \qquad \text{for } \lambda_c > 1.5 \tag{7.16}$$

where

$$\lambda_c = \sqrt{\frac{F_y}{F_e}} \qquad (7.17)$$

The elastic buckling stress (F_e) is the least of the flexural, torsional, or torsional-flexural buckling stresses determined in accordance with Sections C4.1 to C4.3 of the AISI Specification and discussed in Section 7.2 of this book. The resulting column design curve is shown in Figure 7.3, where it is compared with the 1991 AISI Specification column design curve. The design curve was lowered in the 1996 Specification to allow for the effect of geometric imperfection (Ref. 7.3).

To compute the nominal axial strength (P_n) the critical stress (F_n) is multiplied by the effective area (A_e):

$$P_n = A_e F_n \qquad (7.18)$$

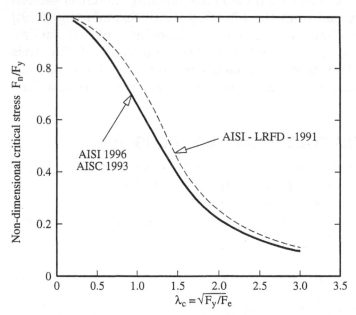

FIGURE 7.3 Compression member design curves.

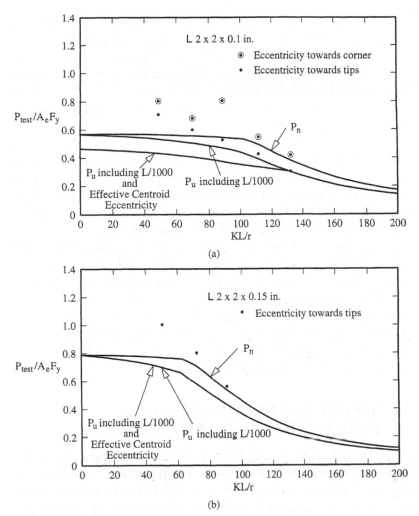

FIGURE 7.4 Pin-ended angle tests (Popovic, Hancock, and Rasmussen): (a) 0.1 in. thick; (b) 0.15 in. thick.

The area A_e is computed at the stress F_n and not the yield stress as for the nominal stub column strength given by Eq. (7.14). The use of the effective area computed at F_n rather than F_y allows for the fact that the section may not be fully stressed when the critical load in compression is reached and hence the effective area is not reduced to its value at yield. The method is called the *unified approach* and is described in detail in Ref. 4.10. It accurately accounts for the interaction of local and overall buckling. It is similar to the method used for beams and described in Chapter 5, where the effective section modulus is computed at the lateral buckling stress.

The AISI Specification requires that angle sections be designed for the axial load (P_u) acting simultaneously with a moment equal to $PL/1000$ applied about the minor principal axis causing compression in the tips of the angle legs. The results of compression tests on a range of DuraGal angles (Ref. 7.4) of 2 in. × 2 in. size and thicknesses 0.10 in. and 0.15 in. are shown in Figures 7.4a, b, respectively. They are compared with P_n and P_u, including $L/1000$ eccentricity. It can be seen from Figure 7.4 that the slender sections (0.10-in. thickness) which fail flexurally-torsionally are best approximated by including the $P_u L/1000$ term, but the stockier sections (0.15 in.) which fail flexurally about the minor principal y-axis are adequately predicted by P_n without the additional eccentricity.

7.5 EFFECT OF LOCAL BUCKLING OF SINGLY-SYMMETRIC SECTIONS

As demonstrated in Figure 7.2, local buckling of a singly symmetric section such as a channel will produce an effective section whose centroid (C_e) is at a different position from the centroid of the gross section. For columns axially loaded between pinned ends, the line of action of the force in the section will move as loading increases and will be eccentric from the line of the applied force. Hence, the section will be subject to eccentric loading, so a bending moment will be applied in addition to the axial force. Section C4 of the AISI Specification requires the designer to allow for the moment resulting from the eccentricity of loading by stating that the axial load passes through the centroid of the effective section. Hence, an initially concentrically loaded pin-ended singly symmetric column must be designed as a member subjected to combined compression and bending as described in Chapter 8.

For fixed-ended columns it has been demonstrated (Ref. 7.5) that the line of action of the applied force moves with the internal line of action of the force which is at the effective centroid and so bending is not induced. Since most columns have some degree of end fixity, design based on loading which is eccentric from the effective centroid may be very conservative.

To investigate the effect of the moving centroid in more detail, both pin-ended and fixed-ended tests were performed on plain (unlipped) channel columns by Young and Rasmussen (Refs. 7.6, 7.7). For the more slender sections tested (P48 with 2-in.-wide flanges), the test results for the fixed-ended tests and for the pin-ended tests are compared with the AISI Specification in Figure 7.5. The fixed-ended tests underwent local and flexural buckling at shorter lengths and torsional-flexural buckling at longer lengths, as shown in Figure 7.5a. The nominal axial strength (P_n) given by Section C4 of the AISI Specification compares closely with the test results. There is no need to include load eccentricity in the design of fixed-ended singly symmetric sections. The pin-ended tests underwent local and flexural buckling at shorter lengths and flexural buckling at longer lengths, as shown in Figure 7.5b. The test results are considerably lower than the nominal axial strength (P_u) given by Section C4 of the AISI Specification. Hence, the effect of loading eccentricity must be included for the pin-ended tests. However, the design curves based on combined compression and bending (as discussed in Chapter 8) are considerably lower than the test results and are thus very conservative.

A similar investigation was performed by Young and Rasmussen (Ref. 7.8) for lipped channel columns. For the more slender sections tested (L48 with 2-in.-wide flanges),

199

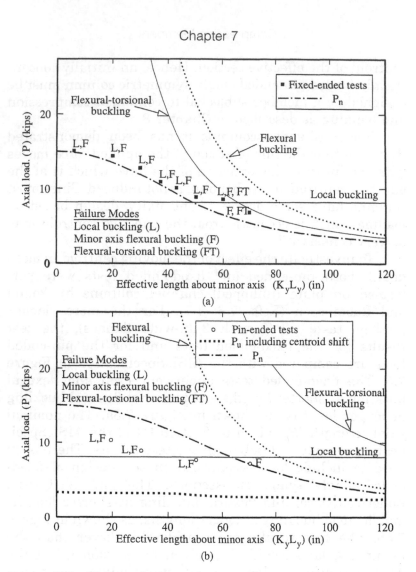

FIGURE 7.5 Unlipped channel tests (Young and Rasmussen): (a) fixed-ended tests; (b) pin-ended tests.

the test results for the fixed-ended tests and pin-ended tests are compared with the AISI Specification in Figure 7.6. The fixed-ended tests underwent local and distortional buckling at shorter lengths and flexural/torsional-flexural

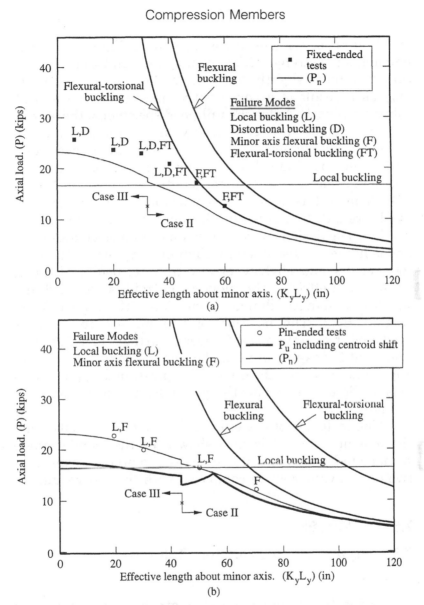

FIGURE 7.6 Lipped channel tests (Young and Rasmussen): (a) fixed-ended tests; (b) pin-ended tests.

buckling at longer lengths, as shown in Figure 7.6a. The nominal axial strength (P_n) given by Section C4 of the AISI Specification is slightly lower than the test results, probably as a result of a conservative prediction of the stub column strength. As for the unlipped specimens, there is no need to include load eccentricity in the design of fixed-ended singly symmetric sections. The pin-ended tests underwent local and flexural buckling at shorter lengths and flexural buckling at longer lengths, as shown in Figure 7.6b. The test results are comparable with the nominal axial strength (P_n) given by Section C4 of the AISI Specification. The design curves based on combined compression and bending (as discussed in Chapter 8) are lower than the test results and slightly erratic at intermediate lengths because the effect of the moving centroid is included and the lipped flanges change from Case III to Case II in Section B4.2. The position of the moving centroid as predicted by Section B4.2 needs further investigation. The need to take account of the moving centroid does not appear to be as significant for the pin-ended lipped channels in Figure 7.6b as for the pin-ended unlipped channels in Figure 7.5b.

The inclusion of the effect of loading eccentricity from the moving centroid is also shown for the slender-angle sections in Figure 7.4a. It can be seen to produce conservative design values when compared with the test results

7.6 EXAMPLES

7.6.1 Square Hollow Section Column

Problem

Determine the design axial strength for the cold-formed square hollow section column shown in Figure 7.7. Assume that the effective length (KL) is 120 in. The nominal yield strength of the material is 50 ksi. The following section and equation numbers refer to the AISI Specification.

202

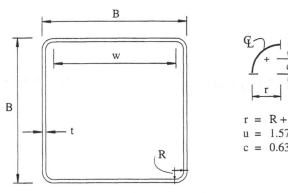

r = R + t/2
u = 1.57r
c = 0.637r

FIGURE 7.7 Square hollow section.

Solution

A. Sectional Properties of Full Section

$$B = 3 \text{ in.} \qquad R = 0.125 \text{ in.} \qquad t = 0.075 \text{ in.}$$

Corner radius $\qquad r = R + \dfrac{t}{2} = 0.1625 \text{ in.}$

Length of arc $\qquad u = 1.57r = 0.255 \text{ in.}$

Distance of centroid $\qquad c = 0.637r = 0.104 \text{ in.}$
from center of radius

$$w = B - 2(R + t) = 2.60 \text{ in.}$$
$$A = 4t(w + u) = 0.8565 \text{ in.}^2$$
$$I_x = 2t\left[\frac{w^3}{12} + w\left(\frac{B - t}{2}\right)^2\right]$$
$$\qquad + 4t\left[u\left(\frac{w}{2} + c\right)^2\right]$$
$$\qquad = 1.205 \text{ in.}^4 = I_y$$
$$r_y = \sqrt{\frac{I_y}{A}} = 1.186 \text{ in.} = r_x$$
$$\frac{KL}{r} = \frac{K_x L_x}{r_x} = \frac{K_y L_y}{r_y} = 101.2$$

203

B. Critical Stress (F_n)

Section C4.1

$$F_e = \sigma_{ex} = \sigma_{ey} = \frac{\pi^2 E}{(KL/r)^2} \qquad \text{(Eq. C4.1-1)}$$

$$= \pi^2 \frac{29500}{101.2^2}$$

$$= 28.4 \text{ ksi}$$

Section C4

$$\lambda_c = \sqrt{\frac{F_y}{F_e}} = \sqrt{\frac{50}{28.4}} = 1.33 \qquad \text{(Eq. C4-4)}$$

Since $\lambda_c < 1.5$, then

$$F_n = 0.658^{\lambda_c^2} F_y = (0.658^{1.769})50 = 0.476 \times 50 = 23.8 \text{ ksi}$$

$$\text{(Eq. C4-2)}$$

C. Effective Area (A_e)
    Section C4
    The effective area (A_e) is computed at the critical stress F_n.
    Section B2.1
    The flange of the SHS is a stiffened element:

$w = B - 2(R + t) = 2.60$ in.

$k = 4$

$f = F_n = 23.8$ ksi

$$\lambda = \frac{1.052}{\sqrt{k}} \left(\frac{w}{t}\right)\sqrt{\frac{f}{E}} \qquad \text{(Eq. B2.1-4)}$$

$$= \frac{1.052}{\sqrt{4}} \left(\frac{2.60}{0.075}\right)\sqrt{\frac{23.8}{29500}}$$

$$= 0.518$$

Since

$\lambda < 0.673, \qquad$ then $b = w \qquad$ (Eq. B2.1-1)

Hence

$$A_e = A = 0.8565 \text{ in.}^2$$

D. Nominal Axial Strength (P_n)

Section C4

$$P_n = A_e F_n = 0.8565 \times 23.8 = 20.4 \text{ kips}$$

LRFD $\phi_c = 0.85$

$$P_d = \phi_c P_n = 17.3 \text{ kips} \qquad\qquad \text{(Eq. C4-1)}$$

ASD $\Omega_c = 1.80$

$$P_d = \frac{P_n}{\Omega_c} = 11.3 \text{ kips}$$

7.6.2 Unlipped Channel Column

Problem

Determine the design axial strength for the channel section shown in Figure 7.8, assuming the channel is loaded concentrically through the centroid of the effective section and the effective lengths in flexure and torsion are 60 in.

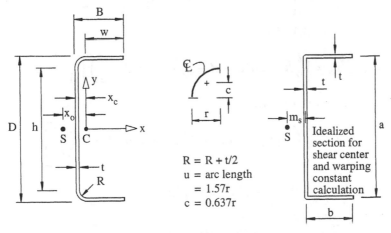

FIGURE 7.8   Unlipped channel section.

The nominal yield stress (F_y) is 36 ksi. The following section and equation numbers refer to the AISI Specification.

Solution

$$K_xL_x = 60 \text{ in.} \quad K_yL_y = 60 \text{ in.} \quad K_tL_t = 60 \text{ in.} \quad F_y = 36 \text{ ksi}$$
$$B = 2 \text{ in.} \qquad D = 6 \text{ in.} \qquad R = 0.125 \text{ in.} \quad t = 0.125 \text{ in.}$$

A. Section Properties of Full Section

$$\text{Corner radius} \qquad r = R + \frac{t}{2} = 0.188 \text{ in.}$$

$$\text{Length of arc} \qquad u = 1.57r = 0.294 \text{ in.}$$

$$\text{Distance of centroid} \qquad c = 0.637r = 0.119 \text{ in.}$$
$$\text{from center of radius}$$

$$w = B - (R + t) = 1.75 \text{ in.}$$
$$h = D - 2(R + t) = 5.5 \text{ in.}$$
$$A = t(2w + h + 2u) = 1.199 \text{ in.}^2$$
$$I_x = 2t\left[u\left(\frac{h}{2}+c\right)^2 + w\left(\frac{D}{2}-\frac{t}{2}\right)^2\right] + t\frac{h^3}{12}$$
$$= 6.114 \text{ in.}^4$$
$$x_c = \left[\frac{t}{2}h + 2u(t + R - c) + 2w\left(B - \frac{w}{2}\right)\right]\frac{t}{A}$$
$$= 0.455 \text{ in.}$$
$$I_y = \left[\left(\frac{t}{2}\right)^2 h + 2u(t + R - c)^2 + 2w\left(B - \frac{w}{2}\right)^2\right]t$$
$$+ 2t\frac{w^3}{12} - Ax_c^2 = 0.422 \text{ in.}^4$$
$$J = \frac{At^2}{3} = 6.243 \times 10^{-3} \text{ in.}^4$$

Shear Center and Warping Constant

$$a = D - t = 5.875 \text{ in.} \qquad b = B - \frac{t}{2} = 1.938 \text{ in.}$$

$$m_s = \frac{3a^2b^2}{a^3 + 6a^2b} = 0.644 \text{ in.}$$

$$x_0 = -\left(m_s - \frac{t}{2}\right) - x_c = -1.036 \text{ in.}$$

$$I_w = a^2b^2t\frac{2a^3b + 3a^2b^2}{12(a^3 + 6a^2b)} = 2.624 \text{ in.}^6$$

$$r_y = \sqrt{\frac{I_y}{A}} = 0.593 \text{ in.} \qquad r_x = \sqrt{\frac{I_x}{A}} = 2.259 \text{ in.}$$

$$r_0 = \sqrt{r_x^2 + r_y^2 + x_0^2} = 2.554 \text{ in.}$$

(Eq. C3.1.2-13)

B. Critical Stress (F_n) According to Section C4

Use Section C3.1.2 to compute σ_{ex}, σ_{ey}, and σ_t:

$$\sigma_{ex} = \frac{\pi^2 E}{(K_x L_x/r_x)^2} = 412.6 \text{ ksi} \qquad \text{(Eq. C3.1.2-8)}$$

$$\sigma_{ey} = \frac{\pi^2 E}{(K_y L_y/r_y)^2} = 28.46 \text{ ksi} \qquad \text{(Eq. C3.1.2-9)}$$

$$\sigma_t = \frac{GJ(1 + \pi^2 EC_w/GJ(K_t L_t)^2)}{Ar_0^2} = 36.16 \text{ ksi}$$

(Eq. C3.1.2-10)

Section C4.2

$$F_{e1} = \sigma_{ey} = 28.46 \text{ ksi}$$

$$F_{e2} = \frac{(\sigma_{ex} + \sigma_t) - \sqrt{(\sigma_{ex} + \sigma_t)^2 - 4[1 - (x_0/r_0)^2]\sigma_{ex}\sigma_t}}{2[1 - (x_0/r_0)^2]}$$

$$= 35.60 \text{ ksi} \hspace{3cm} \text{(Eq. C4.2-1)}$$

$$F_e = \text{lesser of } F_{e1} \text{ and } F_{e2} = 28.46 \text{ ksi}$$

$$\lambda_c = \sqrt{\frac{F_y}{F_e}} = 1.125 \hspace{3cm} \text{(Eq. C4-4)}$$

Since

$$\lambda_c < 1.5, \hspace{1cm} \text{then } F_n = 0.658^{\lambda_c^2} F_y = 21.2 \text{ ksi}$$

$$\text{(Eq. C4-2)}$$

C. Effective Area at Critical Stress (F_n)

Section B2 applied to web of channel

$$k = 4 \hspace{1cm} f = F_n$$

$$\lambda = \frac{1.052}{\sqrt{k}} \left(\frac{h}{t}\right)\sqrt{\frac{f}{E}} = 0.62 \hspace{2cm} \text{(Eq. B2.1-4)}$$

Since

$$\lambda < 0.673, \hspace{1cm} \text{then } b_w = h = 5.5 \text{ in.} \hspace{1cm} \text{(Eq. B2.1-1)}$$

Section B3.1 applied to flanges of channel

$$k = 0.43$$

$$\lambda = \frac{1.052}{\sqrt{k}} \left(\frac{w}{t}\right)\sqrt{\frac{f}{E}} = 0.602 \hspace{2cm} \text{(Eq. B2.1-4)}$$

Since

$$\lambda < 0.673, \hspace{1cm} \text{then } b_f = w = 1.75 \text{ in.} \hspace{1cm} \text{(Eq. B2.1-1)}$$

D. Nominal Axial Strength (P_n)

   Section C4

   $$A_e = t(b_w + 2b_f + 2u) = 1.199 \text{ in.}^2$$
   $$P_n = A_e F_n = 25.41 \text{ kips}$$

   (Eq. C4-1)

   <u>Centroid of Effective Section Under Axial Force Alone</u>

   $$x_{cea} = \left[\frac{t}{2}b_w + 2u(t + R - c) + 2b_f\left(t + R + \frac{b_f}{2}\right)\right]\frac{t}{A_e}$$
   $$= 0.455 \text{ in.}$$

E. Effective Area at Yield Stress (F_y)

   Section B2 applied to web of channel

   $$k = 4 \qquad f = F_y$$

   $$\lambda = \frac{1.052}{\sqrt{k}}\left(\frac{h}{t}\right)\sqrt{\frac{f}{E}} = 0.808$$

   (Eq. B2.1-4)

   $$\rho = \frac{1 - 0.22/\lambda}{\lambda} = 0.90$$

   (Eq. B2.1-3)

   Since

   $$\lambda > 0.673, \qquad \text{then } b_w = \rho h = 4.952 \text{ in.} \qquad \text{(Eq. B2.1-2)}$$

   Section B3.1 applied to flanges of channel

   $$k = 0.43$$

   $$\lambda = \frac{1.052}{\sqrt{k}}\left(\frac{w}{t}\right)\sqrt{\frac{f}{E}} = 0.785$$

   (Eq. B2.1-4)

   $$\rho = \frac{1 - 0.22/\lambda}{\lambda} = 0.917$$

   (Eq. B2.1-3)

   Since

   $$\lambda > 0.673, \qquad \text{then } b_f = \rho w = 1.605 \text{ in.} \qquad \text{(Eq. B2.1-2)}$$

F.   Nominal Section Stub Column Strength (P_{no})

Note that P_{no} is required for Example 8.5.1, Combined Loading.

Section C4

$$A_e = t(b_w + 2b_f + 2u) = 1.094 \text{ in.}^2$$
$$P_{no} = A_e F_y = 39.38 \text{ kips} \qquad \text{(Eq. C4-1)}$$

(Note that $P_n = 25.41$ kips $< P_{no} = 39.38$ kips.)

G.   Design Axial Strength (P_d)

LRFD $\phi_c = 0.85$

$\qquad\qquad P_d = \phi_c P_n = 21.60$ kips

ASD $\Omega_c = 1.80$

$$P_d = \frac{P_n}{\Omega_c} = 14.34 \text{ kips}$$

7.6.3 Lipped Channel Column

Problem

Determine the nominal axial strength (P_n) for the channel section of length 80 in. shown in Figure 7.9, assuming the channel is loaded concentrically through the centroid of the effective section and the effective lengths in flexure and torsion are based on lateral and torsional restraints in the plane of symmetry at mid-height. The nominal yield stress (F_y) is 45 ksi. The following section numbers refer to the AISI Specification.

Solution

$K_x L_x = 80$ in. $K_y L_y = 40$ in. $K_t L_t = 40$ in. $F_y = 45$ ksi

$\qquad B = 3$ in. $\qquad H = 4$ in. $\qquad D = 0.625$ in. $R = 0.125$ in.

$\qquad t = 0.060$ in.

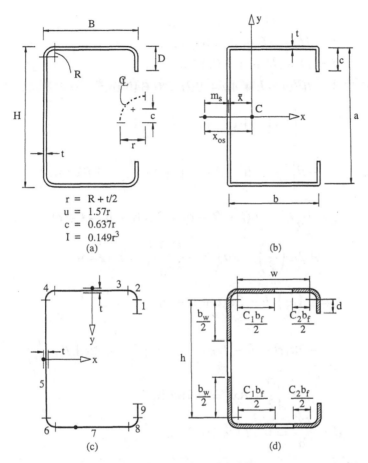

$$r = R + t/2$$
$$u = 1.57r$$
$$c = 0.637r$$
$$I = 0.149r^3$$

(a)

(b)

(c)

(d)

FIGURE 7.9 Lipped channel section: (a) actual section; (b) idealised section; (c) line element model; (d) effective widths.

A. Section Properties of Full Section

$$\text{Corner radius} \quad r = R + \frac{t}{2} = 0.155 \text{ in.}$$

$$\text{Length of arc} \quad u = 1.57 \quad r = 0.243 \text{ in.}$$

$$\begin{array}{c}\text{Distance of centroid} \\ \text{from center of radius}\end{array} \quad c = 0.637 \quad r = 0.099 \text{ in.}$$

211

$$w = B - 2(R+t) = 2.63 \text{ in.}$$
$$h = H - 2(R+t) = 3.63 \text{ in.}$$
$$d = D - (R+t) = 0.44 \text{ in.}$$
$$A = t(2w + h + 4u + 2d) = 0.645 \text{ in.}^2$$

$$I_x = 2t\left[2u\left(\frac{h}{2}+c\right)^2 + w\left(\frac{H}{2}-\frac{t}{2}\right)^2 \right.$$

$$\left. + d\left[\frac{H}{2} - (R+t) - \frac{d}{2}\right]^2 \right] + t\frac{h^3}{12} = 1.812 \text{ in.}^4$$

$$x_c = \left[\frac{t}{2}h + 2u(t+R-c) + 2u(B-R-t+c)\right.$$

$$\left. + 2w\left(\frac{B}{2}\right) + 2d\left(B-\frac{t}{2}\right)\right]\frac{t}{A} = 1.124 \text{ in.}$$

$$I_y = \left[\left(\frac{t}{2}\right)^2 h + 2u(t+R-c)^2 \right.$$

$$\left. + 2u(B-R-t+c)^2 + 2w\left(\frac{B}{2}\right)^2 + 2d\left(B-\frac{t}{2}\right)^2\right]t$$

$$+ 2t\frac{w^3}{12} - Ax_c^2 = 0.792 \text{ in.}^4$$

$$J = \frac{At^2}{3} = 0.774 \times 10^{-3} \text{ in.}^4$$

Shear Center and Warping Constant

$$a = H - t = 3.94 \text{ in.} \qquad b = B - t = 2.94 \text{ in.}$$

$$c = D - \frac{t}{2} = 0.595 \text{ in.}$$

$$m_s = \frac{b(3a^2 b + c(6a^2 - 8c^2))}{a^3 + 6a^2 b + c(8c^2 - 12ac + 6a^2)}$$

$$= 1.493 \text{ in.}$$

$$x_0 = -\left(m_s - \frac{t}{2}\right) - x_c = -2.587 \text{ in.}$$

$$C_w = \frac{a^2b^2t}{12} \left\{ \frac{\begin{array}{c} 2a^3b + 3a^2b^2 + 48c^4 + 112bc^3 + 8ac^3 \\ + 48abc^2 + 12a^2c^2 + 12a^2bc + 6a^3c \end{array}}{6a^2b + (a+2c)^3 - 24ac^2} \right\}$$

$$= 2.951 \text{ in.}^6$$

$$r_y = \sqrt{\frac{I_y}{A}} = 1.109 \text{ in.} \qquad r_x = \sqrt{\frac{I_x}{A}} = 1.677 \text{ in.}$$

$$r_0 = \sqrt{\frac{I_x + I_y}{A} + x_0^2} = 3.28 \text{ in.} \qquad \text{(Eq. C3.1.2-13)}$$

B. Critical Stress (F_n) According to Section C4

Use Section C3.1.2 to compute σ_{ex}, σ_{ey}, and σ_t:

$$\sigma_{ex} = \frac{\pi^2 E}{(K_x L_x / r_x)^2} = 127.9 \text{ ksi} \qquad \text{(Eq. C3.1.2-8)}$$

$$\sigma_{ey} = \frac{\pi^2 E}{(K_y L_y / r_y)^2} = 223.6 \text{ ksi} \qquad \text{(Eq. C3.1.2-9)}$$

$$\sigma_t = \frac{GJ(1 + \pi^2 E C_w / GJ(K_t L_t)^2)}{A r_0^2} = 78.89 \text{ ksi}$$

$$\text{(Eq. C3.1.2-10)}$$

Section C4.2

$$F_{e1} = \sigma_{ey} = 223.6 \text{ ksi}$$

$$F_{e2} = \frac{(\sigma_{ex} + \sigma_t) - \sqrt{(\sigma_{ex} + \sigma_t)^2 - 4[1 - (x_0/r_0)^2]\sigma_{ex}\sigma_t}}{2[1 - (x_0/r_0)^2]}$$

$$= 54.13 \text{ ksi} \qquad \text{(Eq. C4.2-1)}$$

$$F_e = \text{lesser of } F_{e_1} \text{ and } F_{e2} = 54.13 \text{ ksi}$$

$$\lambda_c = \sqrt{\frac{F_y}{F_e}} = 0.912 \qquad \text{(Eq. C4-4)}$$

Since $\lambda_c < 1.5$, then

$$F_n = 0.658^{\lambda_c^2} F_y = 31.78 \text{ ksi} \qquad \text{(Eq. C4-2)}$$

C. **Effective Area at Critical Stress (F_n) as Shown in Figure 7.9d**

Section B2 applied to web of channel

$$k = 4 \qquad f = F_n$$

$$\lambda = \frac{1.052}{\sqrt{k}} \left(\frac{h}{t}\right) \sqrt{\frac{f}{E}} = 1.044 \qquad \text{(Eq. B2.1-4)}$$

$$\rho = \frac{1 - 0.22}{\lambda} = 0.756 \qquad \text{(Eq. B2.1-3)}$$

Since $\lambda > 0.673$, then

$$b_w = \rho h = 2.744 \text{ in.} \qquad \text{(Eq. B2.1-2)}$$

Section B3.1 applied to lips of channel

$$d = 0.44 \text{ in.} \qquad k_u = 0.43$$

$$\lambda = \frac{1.052}{\sqrt{k_u}} \left(\frac{d}{t}\right) \sqrt{\frac{f}{E}} = 0.386 \qquad \text{(Eq. B2.1-4)}$$

Since $\lambda < 0.673$, then

$$d'_s = d = 0.44 \text{ in.} \qquad \text{(Eq. B2.1-1)}$$

$$I_s = \frac{d^3 t}{12} = 0.426 \times 10^{-3} \text{ in.}^4$$

Element 3   Flange flat

$$w = 2.63 \text{ in.}$$

$$\frac{w}{t} = 43.83 \qquad \text{(this value must not exceed 60)}$$

$$\text{(Section B1.1(a))}$$

Section B4.2   Uniformly compressed elements with an edge stiffener (see Figure 7.9d)

$$S = 1.28 \sqrt{\frac{E}{f}} = 39.00 \qquad \text{(Eq. B4-1)}$$

Case I ($w/t \leqslant S/3$) Flange fully effective without stiffener: Not applicable since $w/t > S/3$.

Case II ($S/3 < w/t < S$) Flange fully effective with $I_s \geqslant I_a$

$$I_{a2} = 399t^4\left(\frac{w/t}{S} - \sqrt{\frac{k_u}{4}}\right)^3 = 2.608 \times 10^{-3} \text{ in.}^4 \quad \text{(Eq. B4.2-4)}$$

$n2 = 0.5$

Case III ($w/t \geqslant S$) Flange not fully effective

$$I_{a3} = t^4\left(115\frac{w/t}{S} + 5\right) = 1.740 \times 10^{-3} \text{ in.}^4 \quad \text{(Eq. B4.2-11)}$$

$n3 = 0.333$

Since

$$\frac{w}{t} \geqslant S, \qquad \text{then } I_a = I_{a3} = 1.740 \times 10^{-3} \text{ in.}^4$$
$$n = n3 = 0.333$$

Calculate buckling coefficient (k) and stiffener reduced effective width (d_s).

$$k_a = 5.25 - 5\left(\frac{D}{w}\right) = 4.21 > 4.0 \qquad \text{(Eq. B4.2-8)}$$

Hence

$$k_a = 4.0$$
$$C_2 = \frac{I_s}{I_a} = 0.245 < 1.0 \qquad \text{(Eq. B4.2-5)}$$
$$C_1 = 2 - C_2 = 1.755 \qquad \text{(Eq. B4.2-6)}$$
$$k = C_2^n(k_a - k_u) + k_u = 2.664 \qquad \text{(Eq. B4.2-7)}$$
$$d_s = C_2 d_s' = 0.108 \text{ in.}$$

Section B2.1(a)   Effective width of flange element 3 for strength (see Figure 7.9d)

$$\lambda = \frac{1.052}{\sqrt{k}} \left(\frac{w}{t}\right) \sqrt{\frac{f}{E}} = 0.927 \qquad \text{(Eq. B2.1-4)}$$

$$\rho = \frac{1 - 0.22/\lambda}{\lambda} = 0.823 \qquad \text{(Eq. B2.1-3)}$$

Since $\lambda > 0.673$, then

$$b_f = \rho w = 2.164 \text{ in.} \qquad \text{(Eq. B2.1-2)}$$

D.   Nominal Axial Strength (P_n)

   Section C4

$$A_e = t(b_w + 2b_f + 4u + 2d_s) = 0.496 \text{ in.}^2$$
$$P_n = A_e F_n = 15.75 \text{ kips} \qquad \text{(Eq. C4-1)}$$

Centroid of effective section under axial force alone

$$x_1 = \left[\frac{t}{2}b_w + 2u(t + R - c) + 2u(B - R - t + c)\right]$$
$$= 1.542 \text{ in.}^2$$
$$x_2 = \left[2C_1\frac{b_f}{2}\left(R + t + C_1\frac{b_f}{4}\right)\right.$$
$$\left. +2C_2\frac{b_f}{2}\left(R + t + w - C_2\frac{b_f}{4}\right) + 2d_s\left(B - \frac{t}{2}\right)\right]$$
$$= 6.368 \text{ in.}^2$$
$$x_{cea} = (x_1 + x_2)\frac{t}{A_e} = 0.958 \text{ in.}$$

E.   Effective Area at Yield Stress (F_y)

Section B2 applied to web of channel

$k = 4 \qquad f = F_y$

$$\lambda = \frac{1.052}{\sqrt{k}} \left(\frac{h}{t}\right) \sqrt{\frac{f}{E}} = 1.243 \qquad \text{(Eq. B2.1-4)}$$

$$\rho = \frac{1 - 0.22/\lambda}{\lambda} = 0.662 \qquad \text{(Eq. B2.1-3)}$$

Since $\lambda > 0.673$, then

$$b_w = \rho w = 2.404 \text{ in.} \qquad \text{(Eq. B2.1-2)}$$

Section B3.1 applied to lips of channel

$$d = 0.44 \text{ in.} \qquad k_u = 0.43$$

$$\lambda = \frac{1.052}{\sqrt{k_u}} \left(\frac{d}{t}\right) \sqrt{\frac{f}{E}} = 0.459 \qquad \text{(Eq. B2.1-4)}$$

Since $\lambda < 0.673$, then

$$d'_s = d = 0.44 \text{ in.} \qquad \text{(Eq. B2.1-1)}$$

$$I_s = \frac{d^3 t}{12} = 0.426 \times 10^{-3} \text{ in.}^4$$

Element 3   Flange flat

$w = 2.63 \text{ in.}$

$\dfrac{w}{t} = 43.83 \qquad$ (this value must not exceed 60)

(Clause B1.1(a))

Section B4.2   Uniformly compressed elements with an edge stiffener (see Figure 7.9d)

$$S = 1.28\sqrt{\frac{E}{f}} = 32.77 \qquad\qquad \text{(Eq. B4-1)}$$

Case I $(w/t \leqslant S/3)$ Flange fully effective without stiffener

Not applicable since $\dfrac{w}{t} > \dfrac{S}{3}$.

Case II $(S/3 < w/t < S)$ Flange fully effective with $I_s \geqslant I_a$

$$I_{a2} = 399t^4\left(\frac{w/t}{S} - \sqrt{\frac{k_u}{4}}\right)^3 = 5.322 \times 10^{-3} \text{ in.}^4 \quad \text{(Eq. B4.2-4)}$$

$n2 = 0.5$

Case III $(w/t \geqslant S)$ Flange not fully effective

$$I_{a3} = t^4\left(115\frac{w/t}{S} + 5\right) = 2.058 \times 10^{-3} \text{ in.}^4 \quad \text{(Eq. B4.2-11)}$$

$n3 = 0.333$

Since $\dfrac{w}{t} > S$, then

$$I_a = I_{a3} = 2.058 \times 10^{-3} \text{ in.}^4$$
$$n = n3 = 0.333$$

Calculate buckling coefficient (k) and stiffener-reduced effective width (d_s):

$$k_a = 5.25 - 5\left(\frac{D}{w}\right) = 4.21 > 4.0 \qquad\qquad \text{(Eq. B4.2-8)}$$

Hence,

$k_a = 4.0$

$$C_2 = \frac{I_s}{I_a} = 0.207 < 1.0 \qquad \text{(Eq. B4.2-5)}$$

$C_1 = 2 - C_2 = 1.793$ (Eq. B4.2-6)

$k = C_2^n(k_a - k_u) + k_u = 2.543$ (Eq. B4.2-7)

$d_s = C_2 d'_s = 0.091$ in.   (Eq. B4.2-9)

Section B2.1(a) Effective width of flange element 3 for strength (see Figure 7.9d)

$$\lambda = \frac{1.052}{\sqrt{k}}\left(\frac{w}{t}\right)\sqrt{\frac{f}{E}} = 1.129 \qquad \text{(Eq. B2.1-4)}$$

$$\rho = \frac{1 - 0.22/\lambda}{\lambda} = 0.713 \qquad \text{(Eq. B2.1-3)}$$

Since

$\lambda > 0.673,$ then $b_f = \rho w = 1.875$ in.   (Eq. B2.1-2)

F.   Stub Column Strength (P_{no})

    Section C4

$A_e = t(b_w + 2b_f + 4u + 2d_s) = 0.439$ in.^{2}

$P_{no} = A_e F_y = 19.74$ kips   (Eq. C4-1)

(Note that $P_n = 15.75$ kips $< P_{no} = 19.74$ kips.)

G.   Design Axial Strength (P_d)

    LRFD $\phi_c = 0.85$

                 $P_d = \phi_c P_n = 13.39$ kips

    ASD $\Omega_c = 1.80$

$$P_d = \frac{P_n}{\Omega_c} = 8.75 \text{ kips}$$

REFERENCES

7.1 Hancock, G. J. and Roos, O., Flexural-Torsional Buckling of Storage Rack Columns, Eighth International Specialty Conference on Cold-Formed Steel Structures, St Louis, MO, November 1986.

7.2 Rack Manufacturers Institute, Specification for the Design, Testing and Utilization of Industrial Steel Storage Racks, Part I, June 1997 edition, Charlotte, NC.

7.3 Peköz, T. B. and Sumer, O., Design Provision for Cold-Formed Steel Columns and Beam-Columns, Final report submitted to AISI, Cornell University, September 1992.

7.4 Popovic, D., Hancock, G. J., and Rasmussen, K. J. R., Axial compression tests of DuraGal angles, Journal of Structural Engineering, ASCE, Vol. 125, No. 5, 1999, pp. 515–523.

7.5 Rasmussen, K. J. R. and Hancock, G. J., The flexural behaviour of fixed-ended channel section columns, Thin-walled Structures, Vol. 17, No. 1, 1993, pp. 45–63.

7.6 Young, B. and Rasmussen, K. J. R., Compression Tests of Fixed-ended and Pin-ended Cold-formed Plain Channels, Research Report R714, School of Civil and Mining Engineering, University of Sydney, September 1995.

7.7 Young, B. and Rasmussen, K. J. R., Tests of fixed-ended plain channel columns, Journal of Structural Engineering, ASCE, Vol. 124, No. 2, 1998, pp. 131–139.

7.8 Young, B. and Rasmussen, K. J. R., Design of lipped channel columns, Journal of Structural Engineering, ASCE, Vol. 124, No. 2, 1998, pp. 140–148.

8

Members in Combined Axial Load
and Bending

8.1 COMBINED AXIAL COMPRESSIVE LOAD
AND BENDING: GENERAL

Structural members which are subjected to simultaneous bending and compression are commonly called beam-columns. The bending generally results from one of three sources, depending upon the application of the beam-column. These sources are shown in Figure 8.1a and are described as follows:

(i)   Eccentric axial load of the type usually encountered in columns where the line of action of the axial force is eccentric from the centroid. A typical application is the columns of industrial steel storage racks where the beams apply an eccentric load to the columns as a consequence of the nature of the connections. In the studs in steel-framed housing, particular end connections at the top and bottom plates may produce eccentric loading.

(ii)   Distributed transverse loading normally result- ing from wind forces on sheeting attached to the structural member. A typical application is the purlins in the end spans of industrial buildings where the end wall applies axial load to the purlin and the sheeting applies a distributed load. Another application is the stud walls of steel-framed houses where lateral load from wind and axial load from upper stories or the roof can occur simultaneously.

(iii)   End moments, shears, and axial forces which occur in members of rigid jointed structures such as portal frames, where the members are rigidly connected together using bolted joints.

The combination of concentric axial force and bending moment at a cross section can be converted to an equivalent eccentric axial force as shown in Figure 8.1b. The eccen- tricity of this axial force defines three different situations. For the singly symmetric section shown in Figure 8.1b, these are

(i)   Eccentric compression with bending in the plane of symmetry when $e_x \neq 0$ and $e_y = 0$
(ii)   Eccentric compression with bending about the axis of symmetry when $e_x = 0$ and $e_y \neq 0$
(iii)   Biaxial bending when $e_x \neq 0$ and $e_y \neq 0$

In the AISI Specification, Section C5 specifies the accepta- ble combinations of action effects (force, moment) for members subject to combined bending and compression. Section C5 uses the design capacity for beams given in Section C3 as well as those for columns given in Section C4. The resistance factors (ϕ_b, ϕ_c) and safety factors (Ω_b, Ω_c) for bending and compression remain the same as when the bending or compression is considered alone.

(i) Eccentric axial load

(ii) Transverse and axial load

(iii) End moments, shears and axial load

(a)

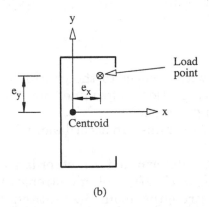

(b)

FIGURE 8.1   Combined compression and bending: (a) member loads; (b) cross-section loads.

8.2 INTERACTION EQUATIONS FOR COMBINED AXIAL COMPRESSIVE LOAD AND BENDING

The rules governing combined actions are based on the use of linear interaction equations expressed in terms of axial force and moment. Two sets of equations are provided in the AISI Specification, one set for ASD and one set for

223

LRFD. The set for ASD is given in Section C5.2.1 of the AISI Specification as

When $\Omega_c P/P_n \leqslant 0.15$,

$$\frac{\Omega_c P}{P_{no}} + \frac{\Omega_b M_x}{M_{nx}} + \frac{\Omega_b M_y}{M_{ny}} \leqslant 1.0 \tag{8.1}$$

When $\Omega_c P/P_n > 0.15$,

(a) $$\frac{\Omega_c P}{P_n} + \frac{\Omega_b C_{mx} M_x}{M_{nx}\alpha_x} + \frac{\Omega_b C_{my} M_y}{M_{ny}\alpha_y} \leqslant 1.0 \tag{8.2}$$

(b) $$\frac{\Omega_c P}{P_{no}} + \frac{\Omega_b M_x}{M_{nx}} + \frac{\Omega_b M_y}{M_{ny}} \leqslant 1.0 \tag{8.3}$$

where P = nominal axial load
M_x, M_y = nominal moments with respect to the centroidal axes of the effective section determined for the required axial strength alone
P_n = nominal axial strength determined in accordance with Section C4
P_{no} = nominal axial strength determined in accordance with Section C4 with $F_n = F_y$ (stub column strength)
M_{nx}, M_{ny} = nominal flexural strengths about the centroidal axes determined in accordance with Section C3
α_x, α_y = moment amplification factors as discussed below
C_{mx}, C_{my} = coefficients for unequal end moments as discussed below

The set for LRFD is given in Section C5.2.2.
When $P_u/\phi_c P_n \leqslant 0.15$,

$$\frac{P_u}{\phi_c P_n} + \frac{M_{ux}}{\phi_b M_{nx}} + \frac{M_{uy}}{\phi_b M_{ny}} \leqslant 1.0 \tag{8.4}$$

When $P_u/\phi_c P_n > 0.15$,

(a) $\dfrac{P_u}{\phi_c P_n} + \dfrac{C_{mx}M_{ux}}{\phi_b M_{nx}\alpha_x} + \dfrac{C_{my}M_{uy}}{\phi_b M_{ny}\alpha_y} \leqslant 1.0$ (8.5)

(b) $\dfrac{P_u}{\phi_c P_{no}} + \dfrac{M_{ux}}{\phi_b M_{nx}} + \dfrac{M_{uy}}{\phi_b M_{ny}} \leqslant 1.0$ (8.6)

where P_u = required axial strength

M_{ux}, M_{uy} = required flexural strengths with respect to the centroidal axes of the effective section determined for the required axial strength alone

P_n = nominal axial strength determined in accordance with Section C4

P_{no} = nominal axial strength determined in accordance with Section C4 with $F_n = F_y$ (stub column strength)

M_{bx}, M_{by} = nominal flexural strength about the centroidal x- and y-axes, respectively, determined in accordance with Section C3

C_{mx}, C_{my} = coefficients for unequal end moments as discussed below

α_x, α_y = moment amplification factors as discussed below

The linear interaction equations expressed as Eqs. (8.1) to (8.3) for ASD or Eqs. (8.4) to (8.6) for LRFD allow for the limit states of in-plane section strength failure, in-plane member strength including in-plane buckling, and out-of-plane member strength including lateral buckling. A detailed discussion of these limit states is given in Ref. 8.1.

    The coefficients C_{mx}, C_{my} allow for unequal end moments in beam-columns. They are set out for members free to sway in Figure 8.2a and members braced against sway in Figure 8.2b.

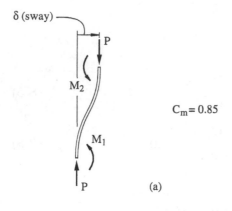

$C_m = 0.85$

(a)

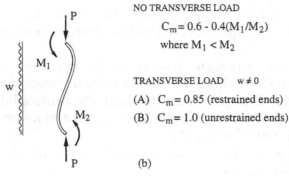

NO TRANSVERSE LOAD

$C_m = 0.6 - 0.4(M_1/M_2)$

where $M_1 < M_2$

TRANSVERSE LOAD $\quad w \neq 0$

(A) $C_m = 0.85$ (restrained ends)

(B) $C_m = 1.0$ (unrestrained ends)

(b)

FIGURE 8.2 $\quad C_m$ factors: (a) free to sway; (b) braced against sway.

The moment amplification factors for ASD are defined by

$$\alpha_x = 1 - \frac{\Omega_c P}{P_{Ex}} \tag{8.7}$$

$$\alpha_y = 1 - \frac{\Omega_c P}{P_{Ey}} \tag{8.8}$$

where

$$P_{Ex} = \frac{\pi^2 E I_x}{(K_x L_x)^2} \tag{8.9}$$

and

$$P_{Ey} = \frac{\pi^2 EI_y}{(K_y L_y)^2} \qquad (8.10)$$

The moment amplification factors for the LRFD method are defined by

$$\alpha_x = 1 - \frac{P_u}{P_{Ex}} \qquad (8.11)$$

$$\alpha_y = 1 - \frac{P_u}{P_{Ey}} \qquad (8.12)$$

where P_{Ex}, P_{Ey} are given by Eqs. (8.9) and (8.10).

The values of I_x, I_y are the second moments of area of the full, unreduced cross section about the x-, y-axis, respectively. The values of $K_x L_x$, $K_y L_y$ are the effective lengths for buckling about the x-, y-axis, respectively. They may be greater than the member length for members in sway frames. The moment amplification factors (α_x, α_y) increase the first-order elastic moments (M_x, M_y) to second-order elastic moments (M_x/α_x, M_y/α_y). The values of C_{mx}/α_{nx} and C_{my}/α_{ny} should never be less than 1.0 since the second-order moment must always be greater than or equal to the first-order moment.

8.3 SINGLY SYMMETRIC SECTIONS UNDER COMBINED AXIAL COMPRESSIVE LOAD AND BENDING

The behavior of singly symmetric sections depends on whether the sections are bent in the plane of symmetry, about the axis of symmetry, or both.

8.3.1 Sections Bent in a Plane of Symmetry

Singly symmetric sections bent in the plane of symmetry may fail by either of two ways:

Chapter 8

(a) Deflecting gradually in the plane of symmetry without twisting, followed by yielding or local buckling at the location of maximum moment. This mode is denoted by the curves labeled "Flexural Yielding" in Figure 8.3, which has been developed from the study described in Ref. 8.2.

(b) Gradual flexural bending in the plane of symmetry, but when the load reaches a critical value, the member will suddenly buckle by torsional-flexural buckling. This mode is denoted by the curve labeled "Torsional-Flexural Buckling" in Figure 8.3.

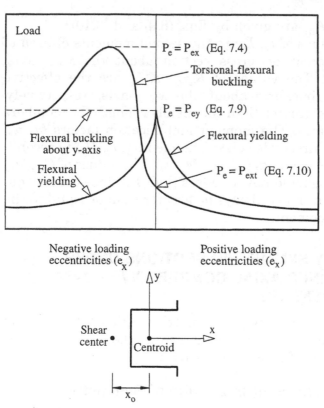

FIGURE 8.3 Modes of failure for eccentric loading in plane of single symmetry.

The type of failure depends upon the shape and dimensions of the cross section, the column length (see Figure 3.4), or the load eccentricity (see Figures 3.3 and 3.4). As demonstrated in Figure 8.3, the mode described in (a) may occur for large negative eccentricities, and the mode described in (b) may occur for large positive eccentricities, even though the column section and length remain unaltered.

The types of failure described in (a) and (b) are both accurately accounted for by using the first and third terms of the linear interaction equations (when the x-axis is the axis of symmetry) or the first and second terms (when the y-axis is the axis of symmetry). The determination of the torsional-flexural buckling moment (M_e) for a section symmetric about the x-axis and bent in the plane of symmetry, as in Figure 8.3, can be calculated by using Eq. (5.11).

A series of tests by S. Wang on cold-formed hat sections with the lips turned outward has been reported by Jang and Chen (Ref. 8.3). The test results of the columns are set out in Figure 8.4 in the same format as Ref. 8.3. The columns were tested between simple supports (distance L apart) with the ends prevented from warping. Consequently the effective lengths ($K_x L_x$, $K_y L_y$, and $K_t L_t$) used in the calculations have been taken as L, L, and $L/2$, respectively, since the warping restraint effectively halves the torsional effective length. The values of F_e computed using Section C4.2 agree closely with the test results when elastic buckling occurs. The shape of the design curves produced by Section C5.2 of the AISI Specification for eccentric loading with positive eccentricity accurately reflects the test results. For negative eccentricity, the design curves are accurate for the shorter columns ($K_x L_x/r_x = 35$) where flexural yielding predominates over torsional-flexural buckling. However, for slender columns loaded with negative eccentricity, the design curves are conservative since the simple linear interaction formula

229

does not accurately predict the beneficial interaction of pure compression and bending in the torsional-flexural mode, as shown by the curve labeled "Torsional-Flexural Buckling" in Figure 8.3. The interaction formula actually predicts an adverse interaction in this case and so the design curves drop, as shown in Figure 8.4. However, it is unlikely that a designer could ever take advantage of the increase in the torsional-flexural buckling load at small negative eccentricities, even if the design curves accurately reflected this phenomenon, since the designer would need to be extremely confident that the column was always loaded with a negative eccentricity and never with a positive eccentricity. Consequently, the conservatism of

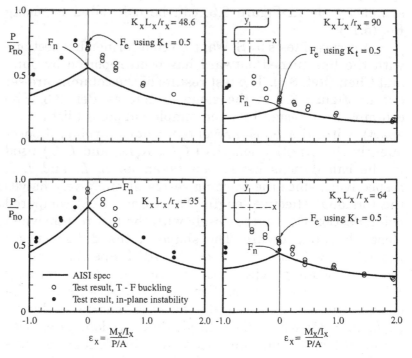

FIGURE 8.4   Comparison of Section C5.2.2 with test results of hat section.

the design curves for small negative eccentricities is unlikely to be important.

8.3.2 Sections Bent About an Axis of Symmetry

The case of bending about the axis of symmetry is an important case in the design of upright columns of industrial storage racks where the pallet beams produce bending about the axis of symmetry in the upright columns. It is also an important case in the design of purlins in the end bays of industrial buildings where the lateral load on the purlin produces bending about the plane of symmetry in addition to axial load from the end wall of the building. The Rack Manufacturers Institute Specification (Ref. 8.4) gives a design approach for this situation, and this has been incorporated into the AISI Specification.

For eccentric compression with the bending about the axis of symmetry, the response of the column involves biaxial bending and twisting even if the column contains no geometric imperfections. The approximate design approach adopted in the AISI Specification uses the conventional beam-column interaction formulae given in Section 8.2, with the buckling stress for concentric axial compression of a singly symmetric section based on Eq. (7.10) and the buckling stress for pure bending about the axis of symmetry based on Eq. (5.9). The experimental justification of this method is given in Refs. 8.5 and 8.6. This justification includes the use of the linear interaction formulae [Eqs. (8.1)–(8.6)] for sections with slender elements.

8.4 COMBINED AXIAL TENSILE LOAD AND BENDING

The design rules for members subject to combined axial tensile load and bending are new, having been added for the first time to the most recent edition of the AISI Specification. They are included in Section C5.1. The design rules consist of two checks. These are a lateral buckling check, given by Eqs. (8.12) and (8.14), and a section capacity in tension check, given by Eqs. (8.11) and (8.13).

For ASD

$$\frac{\Omega_b M_x}{M_{nxt}} + \frac{\Omega_b M_y}{M_{nyt}} + \frac{\Omega_t T}{T_n} \leqslant 1.0 \tag{8.11}$$

$$\frac{\Omega_b M_x}{M_{nx}} + \frac{\Omega_b M_y}{M_{ny}} - \frac{\Omega_t T}{T_n} \leqslant 1.0 \tag{8.12}$$

For LRFD

$$\frac{T_u}{\phi_t T_n} + \frac{M_{ux}}{\phi_b M_{nxt}} + \frac{M_{uy}}{\phi_b M_{nyt}} \leqslant 1.0 \tag{8.13}$$

$$\frac{M_{ux}}{\phi_b M_{nx}} + \frac{M_{uy}}{\phi_b M_{ny}} - \frac{T_u}{\phi_t T_n} \leqslant 1.0 \tag{8.14}$$

where M_{nx}, M_{ny} = nominal flexural strength about the x- and y-axis, respectively, of the effective section as given in Section C3.1

M_{nxt}, M_{nyt} = nominal flexural strength of the full section about the x- and y-axis, respectively, and equal to the section modulus of the full section (S_{ft}) for the extreme tension fiber multiplied by the yield stress (F_y)

The nominal tensile strength of a member (T_n) is specified in Section C2 of the AISI Specification as the lesser of:

$$T_n = A_g F_y \tag{8.15}$$

and

$$T_n = A_n F_u \tag{8.16}$$

where A_g = gross area of the cross section
A_n = net area of the cross section

The values of Ω_t and ϕ_t to be used with Eq. (8.15) are 1.67 and 0.90, respectively, and reflect yielding of the cross section. The values of Ω_t and ϕ_t to be used with Eq. (8.16) are 2.00 and 0.75, respectively, and reflect fracture away from a connection. Equation (8.16) was added to the AISI Specification in 1999 as part of Supplement No. 1.

When bending is combined with axial tension, the effect of bending on the tension section capacity is accounted for by the interaction in Eqs. (8.11) and (8.13), which lowers the design axial tensile force when bending is included. The bending terms in Eqs. (8.11) and (8.13) are based on yield in the extreme tension fiber of the section and therefore use the section modulus of the full unreduced section based on the extreme tension fiber. The effect of bending on the lateral buckling capacity is accounted for by Eqs. (8.12) and (8.14), which increases the design bending moment by subtracting the tension term from the bending terms. Care must be taken with terms of this type to ensure that a design situation cannot occur where the tension may go to zero, although the bending moment remains nonzero.

8.5 EXAMPLES

8.5.1 Unlipped Channel Section Beam-Column Bent in Plane of Symmetry

Problem

Calculate the design axial compressive load in the channel shown in Figure 7.8, assuming that the channel is loaded with an axial force on the line of the x-axis at a point in line with the flange tips. The following section numbers refer to the AISI Specification.

233

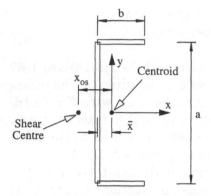

FIGURE 8.5 Square corner model of unlipped channel.

Solution

A. Section Property (j) for Torsional-Flexural Buckling

From Figure 8.5 and Example 7.6.2A:

$$a = 5.875 \text{ in.} \qquad b = 1.938 \text{ in.} \qquad t = 0.125 \text{ in.}$$

$$\bar{x} = \frac{b^2}{a + 2b} = 0.385 \text{ in.}$$

$$x_{os} = -\bar{x} - \frac{3b^2}{a + 6b} = -1.029 \text{ in.}$$

$$I_{ys} = \frac{2b^3 t}{12} + 2bt\left(\frac{b}{2} - \bar{x}\right)^2 + at\bar{x}^2 = 0.425 \text{ in.}^4$$

Note that the distances (x_0 and $\bar{x}$) and the value of I_y have all been computed for the square corner model in Figure 8.5.

$$\beta_w = -\frac{t\bar{x}a^3}{12} - t\bar{x}^3 a = -0.855 \text{ in.}^5$$

$$\beta_f = \tfrac{1}{2}t[(b - \bar{x})^4 - \bar{x}^4] + \tfrac{1}{4}a^2 t[(b - \bar{x})^2 - \bar{x}^2] = 2.801 \text{ in.}^5$$

$$j = \frac{\beta_w + \beta_f}{2I_{ys}} - x_{os} = 3.316 \text{ in.}$$

B. **Elastic Critical Stress for Torsional-Flexural Buckling due to Bending in Plane of Symmetry**

Section C3.1.2.1 (1999 Supplement No. 1)
Use σ_{ex}, σ_t, r_0 from Example 7.6.2: $\sigma_{ex} = 412.6\,\text{ksi}$, $\sigma_t = 36.16\,\text{ksi}$, $r_0 = 2.554\,\text{in}$.

$C_s = -1$ tension on shear center side of centroid

$C_{TF} = 1.0$ uniform moment

$$S_f = \frac{I_y}{B - x_c} = 0.273 \text{ in.}^3$$

$$F_e = \frac{C_s A \sigma_{ex}[\, j + C_s \sqrt{j^2 + r_0^2(\sigma_t/\sigma_{ex})}\,]}{C_{TF} S_f} = 154.2 \text{ ksi}$$

$$\text{(Eq.C3.1.2.1-6)}$$

Since $F_e \geqslant 2.78 F_y$, then

$$F_c = F_y = 36 \text{ ksi} \qquad \text{(Eq. C3.1.2.1-2)}$$

C. **Effective Section Modulus (S_{cy}) and Nominal Flexural Strength (M_{ny}) at a Stress F_c**

For the loading eccentricity causing compression at the flange tips and tension in the web under the bending moment as shown in Figure 8.6, the effective widths of the flanges under bending need to be calculated.
Section B3.2 applied to flanges of channel

$k = 0.43 \qquad f = F_c$

$$\lambda = \frac{1.052}{\sqrt{k}} \left(\frac{w}{t}\right) \sqrt{\frac{f}{E}} = 0.785 \qquad \text{(Eq. B2.1-4)}$$

$$\rho = \frac{1 - 0.22/\lambda}{\lambda} = 0.917 \qquad \text{(Eq. B2.1-3)}$$

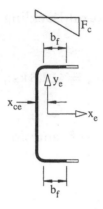

FIGURE 8.6   Effective section for bending in plane of symmetry.

Since $\lambda > 0.673$, then

$$b_f = \rho w = 1.605 \text{ in.}$$

$$A_e = t(2b_f + h + 2u) = 1.162 \text{ in.}^2$$

$$x_{ce} = \left[\frac{t}{2}h + 2u(t + R - c) + 2b_f\left(t + R + \frac{b_f}{2}\right)\right]\frac{t}{A_e}$$

$$= 0.409 \text{ in.}$$

$$I_{ye} = \left[\left(\frac{t}{2}\right)^2 h + 2u(t + R - c)^2 + 2b_f\left(t + R + \frac{b_f}{2}\right)^2\right]t$$

$$+ 2t\frac{b_f^3}{12} - A_e x_{ce}^2 = 0.341 \text{ in.}^4$$

$$S_{cy} = \frac{I_{ye}}{t + R + b_f - x_{ce}} = 0.235 \text{ in.}^3$$

$$M_{ny} = S_{cy}F_c = 8.476 \text{ kip-in.} \qquad \text{(Eq. C3.1.2.1-1)}$$

D.   Combined Axial Compressive Load and Bending

   Section C5   For uniform moment

   $$C_{my} = C_m = 1.0$$

Centroid of effective section under axial force alone from Section 7.6.2D:

$$x_{cea} = 0.455 \text{ in.}$$

Use P_{no}, P_n from Example 7.6.2: $P_{no} = 39.38 \text{ kips}$, $P_n = 25.41 \text{ kips}$.

Section C5.2.1 ASD method

$$\Omega_c = 1.80 \qquad \Omega_b = 1.67$$

Try

$$P = 2.40 \text{ kips}$$
$$M_y = P(B - x_{cea}) = 3.71 \text{ kip-in.}$$
$$A = 1.199 \text{ in.}^2 \qquad\qquad \text{(Example 7.6.2)}$$
$$\sigma_{ey} = 28.46 \text{ ksi}$$
$$P_{Ey} = A\sigma_{ey} = 34.11 \text{ kips}$$
$$\alpha_y = 1 - \frac{\Omega_c P}{P_{Ey}} = 0.873$$

$$\frac{\Omega_c P}{P_n} + \frac{\Omega_b C_{my} M_y}{M_{ny}\alpha_y} = \frac{1.80 \times 2.4}{25.41} + \frac{1.67 \times 1 \times 3.71}{8.476 \times 0.873} = 1.00$$

(Eq. C5.2.1-1)

$$\frac{\Omega_c P}{P_{no}} + \frac{\Omega_b M_y}{M_{ny}} = \frac{1.80 \times 2.4}{39.38} + \frac{1.67 \times 3.71}{8.476} = 0.84$$

(Eq. C5.2.1-2)

Hence, the nominal axial compressive load applied eccentrically at the flange tips is

$$P = 2.40 \text{ kips}$$

Section C5.5.2 LRFD method

$$\phi_c = 0.85 \qquad \phi_b = 0.90$$

Try

$$P_u = 3.65 \text{ kips}$$

$$M_{uy} = P_u(B - x_{cea}) = 5.641 \text{ kip-in.}$$

$$P_{Ey} = A\sigma_{ey} = 34.11 \text{ kips}$$

$$\alpha_y = 1 - \frac{P_y}{P_{Ey}} = 0.893 \qquad \text{(Eq. C5.2.2-5)}$$

$$\frac{P_u}{\phi_c P_n} + \frac{C_{my}M_{uy}}{\phi_b M_{ny}\alpha_y} = \frac{3.65}{0.85 \times 25.41} + \frac{1 \times 5.641}{0.90 \times 8.476 \times 0.893}$$

$$= 0.997 < 1.0$$

$$\text{(Eq. C5.2.2-1)}$$

$$\frac{P_u}{\phi_c P_{no}} + \frac{M_{uy}}{\phi_b M_n} = \frac{3.65}{0.85 \times 39.38} + \frac{5.641}{0.90 \times 8.476}$$

$$= 0.849 < 1.0$$

$$\text{(Eq. C5.2.2-2)}$$

Hence, the required axial strength for a load applied eccentrically at the flange tip is

$$P_u = 3.65 \text{ kips}$$

8.5.2 Unlipped Channel Section Beam-Column Bent About Plane of Symmetry

Problem

Calculate the design axial compressive load in the channel shown in Figure 7.8, assuming that the channel is loaded with an axial force on the intersection of the y-axis with one flange. The following section numbers refer to the AISI Specification.

Solution

A. Elastic Critical Stress for Flexural-Torsional Buckling due to Bending about Plane of Symmetry

Section C3.1.2.1 (1999 Supplement No. 1)

$C_b = 1.0$ uniform moment

$$S_f = \frac{I_x}{D/2} = 2.038 \text{ in.}^3$$

$$F_e = \frac{C_b A r_0}{S_f} \sqrt{\sigma_{ey}\sigma_t} = 48.2 \text{ ksi} \qquad \text{(Eq. C3.1.2.1-5)}$$

Since $2.78F_y > F_e > 0.56F_y$,

$$F_c = \frac{10}{9} F_y \left[1 - \frac{10F_y}{36F_e}\right] = 31.67 \text{ ksi} \qquad \text{(Eq. C3.1.2.1-3)}$$

B. Effective Section Modulus (S_{cx}) and Nominal Flexural Strength (M_{nx}) at a Stress F_c

For the loading eccentricity causing uniform compression on the top flange and bending in the web as shown in Figure 8.7, the effective widths of the compression flange under bending need to be calculated.

Section B3.1 applied to compression flange of channel

$k = 0.43 \qquad f = F_c$

$$\lambda = \frac{1.052}{\sqrt{k}} \left(\frac{w}{t}\right)\sqrt{\frac{f}{E}} = 0.736 \qquad \text{(Eq. B2.1-4)}$$

$$\rho = \frac{1 - 0.22/\lambda}{\lambda} = 0.953 \qquad \text{(Eq. B2.1-3)}$$

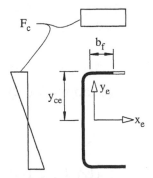

FIGURE 8.7   Effective section for bending about axis of symmetry.

Since $\lambda > 0.679$, then

$$b_f = \rho w = 1.667 \text{ in.} \qquad\qquad (\text{Eq.B2.1-2})$$

$$A_e = t(w + b_f + h + 2u) = 1.188 \text{ in.}^2$$

$$
\begin{aligned}
y_{ce} = &\left[\frac{t}{2}b_f + u(t + R - c) + \frac{D}{2}h \right. \\
&\left. + u(D - t - R + c) + w\left(D - \frac{t}{2}\right)\right]\frac{t}{A_e}
\end{aligned}
$$

$$= 3.026 \text{ in.}$$

$$
\begin{aligned}
I_{xe} = &\left[\left(\frac{t}{2}\right)^2 b_f + u(t + R - c)^2 + \left(\frac{D}{2}\right)^2 h \right. \\
&\left. + u(D - t - R + c)^2 + w\left(D - \frac{t}{2}\right)^2\right]t \\
&+ t\frac{h^3}{12} - A_e y_{ce}^2 = 6.024 \text{ in.}^4
\end{aligned}
$$

$$S_{cx} = \frac{I_{xe}}{y_{ce}} = 1.991 \text{ in.}^3$$

$$M_{nx} = S_{cx}F_c = 63.05 \text{ kip-in.}$$

C.   Combined Axial Compressive Load and Bending

Section C5   For uniform moment

$$C_{mx} = C_m = 1.0$$

Section C5.2.1   ASD method

$$\Omega_c = 1.80 \qquad \Omega_b = 1.67$$

Try

$$P = 6.6 \text{ kips}$$

$$M_x = \frac{PD}{2} = 19.8 \text{ kip-in.}$$

$$A = 1.199 \text{ in.}^2 \qquad\qquad (\text{Example 7.6.2})$$

$$\sigma_{ex} = 412.6 \text{ ksi}$$

$$P_{Ex} = A\sigma_{ex} = 494.5 \text{ kips}$$

$$\alpha_x = 1 - \frac{\Omega_c P}{P_{Ex}} = 0.976 \qquad\qquad (\text{Eq. C5.2.1-4})$$

$$\frac{\Omega_c P}{P_n} + \frac{\Omega_b C_{mx} M_x}{M_{nx} \alpha_x} = \frac{1.80 \times 6.6}{25.41} + \frac{1.67 \times 1 \times 19.8}{63.05 \times 0.976} = 1.00$$

(Eq. C5.2.1-1)

$$\frac{\Omega_c P}{P_{no}} + \frac{\Omega_b M_x}{M_{nx}} = \frac{1.80 \times 6.6}{39.38} + \frac{1.67 \times 19.8}{63.05} = 0.83$$

(Eq. C5.2.1-2)

Hence, the nominal axial compressive load applied eccentrically on one flange is

$$P = 6.6 \text{ kips}$$

Section C5.2.2   LRFD method

$$\phi_c = 0.85 \qquad \phi_b = 0.90$$

Try

$$P_u = 9.9 \text{ kips}$$

$$M_{ux} = P_u \frac{D}{2} = 29.7 \text{ kip-in.}$$

$$P_{Ex} = A\sigma_{ex} = 494.5 \text{ kips}$$

$$\alpha_x = 1 - \frac{P_u}{P_{Ex}} = 0.980 \qquad \text{(Eq. C5.2.2-5)}$$

$$\frac{P_u}{\phi_c P_n} + \frac{C_{mx} M_{ux}}{\phi_b M_{nx} \alpha_x} = \frac{9.9}{0.85 \times 25.41} + \frac{1 \times 29.7}{0.90 \times 63.05 \times 0.98}$$

$$= 0.992 < 1.0 \qquad \text{(Eq. C5.2.2-1)}$$

$$\frac{P_u}{\phi_c P_{no}} + \frac{M_{ux}}{\phi_b M_{nx}} = \frac{9.9}{0.85 \times 39.38} + \frac{29.7}{0.90 \times 63.05}$$

$$= 0.819 < 1.0 \qquad \text{(Eq. C5.2.2-2)}$$

Hence, the required axial strength for a load applied eccentrically on one flange is

$$P_u = 9.9 \text{ kips}$$

8.5.3 Lipped Channel Section Beam-Column Bent in Plane of Symmetry

Problem

Calculate the design axial compressive load in the lipped channel shown in Figure 7.9, assuming that the channel is loaded with an axial force on the intersection of the x-axis with the web outer edge. The following section numbers refer to the AISI Specification.

Solution

A. Section Property (j) for Torsional-Flexural Buckling
From Figure 7.9b and Example 7.6.3,

$$a = 3.94 \text{ in.} \qquad b = 2.94 \text{ in.} \qquad c = 0.595 \text{ in.}$$
$$t = 0.60 \text{ in.}$$

$$m_s = \frac{b(3a^2b + c(6a^2 - 8c^2))}{a^3 + 6a^2b + c(8c^2 - 12ac + 6a^2)} = 1.493 \text{ in.}$$

$$\bar{x} = \frac{b(b + 2c)}{a + 2b + 2c} = 1.103 \text{ in.}$$

$$x_{os} = -m_s - \bar{x} = 2.596 \text{ in.}$$

$$I_{ys} = \frac{2b^3t}{12} + 2bt\left(\frac{b}{2} - \bar{x}\right)^2 + at\bar{x}^2 + 2ct(b - \bar{x})^2 = 0.83 \text{ in.}^4$$

Note that the distances x_{os} and $\bar{x}$ and the value of I_{ys} have all been computed for the square corner model in Figure 7.9b.

$$\beta_w = -\frac{t\bar{x}a^3}{12} - t\bar{x}^3a = -0.654 \text{ in.}^5$$

$$\beta_f = \frac{1}{2}t[(b - \bar{x})^4 - \bar{x}^4] + \frac{1}{4}a^2t[(b - \bar{x})^2 - \bar{x}^2] = 0.80 \text{ in.}^5$$

$$\beta_1 = 2ct(b - \bar{x})^3 + \frac{2}{3}t(b - \bar{x})\left[\left(\frac{a}{2}\right)^3 - \left(\frac{a}{2} - c\right)^3\right]$$

$$= 0.814 \text{ in.}^5$$

$$j = \frac{\beta_w + \beta_f + \beta_1}{2I_{ys}} - x_{os} = 3.174 \text{ in.}$$

B. Elastic Critical Stress for Torsional-Flexural Buckling
    due to Bending in Plane of Symmetry
    Section C3.1.2.1 (1999 Supplement No. 1)
    Use σ_{ex}, σ_t, r_0 from Example 7.6.3: $\sigma_{ex} = 127.9\,\text{ksi}$,
$\sigma_t = 78.89\,\text{ksi}$, $r_0 = 3.28\,\text{in}$.

$\quad C_s = +1 \qquad$ compression on shear center side of
$\qquad\qquad\qquad$ centroid

$\quad C_{TF} = 1.0 \qquad$ uniform moment

$\quad S_f = \dfrac{I_y}{x_c} = 0.705 \text{ in.}^2$

$\quad F_e = \dfrac{C_s A \sigma_{ex}\left[j + C_s\sqrt{j^2 + r_0^2(\sigma_t/\sigma_{ex})}\right]}{C_{TF}S_f} = 849 \text{ ksi}$

Since $F_e \geqslant 2.78 F_y$, then

$\quad F_c = F_y = 45 \text{ ksi} \qquad\qquad\qquad$ (Eq. C3.1.2.1-2)

C. Effective Section Modulus (S_{cy}) and Nominal Flexural
    Strength (M_{ny}) at a Stress F_c

For the loading eccentricity causing tension in the
lips and compression in the web under the bending component, the effective widths of the flanges under bending
may need to be calculated. First the effective width of the
web is calculated in compression. The section is shown in
Figure 8.8.
Section B2.1 applied to web of channel

$\quad k = 4.0 \qquad f = F_c$

$\quad \lambda = \dfrac{1.052}{\sqrt{k}}\left(\dfrac{w}{t}\right)\sqrt{\dfrac{f}{E}} = 1.243 \qquad\qquad$ (Eq. B2.1-4)

$\quad \rho = \dfrac{1 - 0.22/\lambda}{\lambda} = 0.662 \qquad\qquad$ (Eq. B2.1-3)

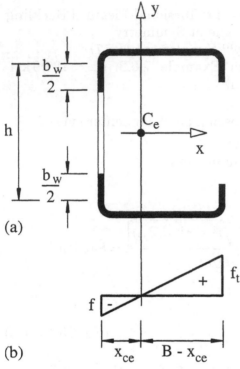

(a)

(b)

FIGURE 8.8 Effective section for bending in plane of symmetry (web in compression): (a) effective section; (b) bending stresses.

Since $\lambda > 0.673$, then

$$b_w = \rho h = 2.404 \text{ in.} \qquad \text{(Eq.B2.1-2)}$$

$$A_e = t(2w + b_w + 4u + 2d) = 0.571 \text{ in.}^2$$

$$x_{ce} = \left[\frac{t}{2} b_w + 2w\left(\frac{B}{2}\right) + 2u(t + R - c) \right.$$

$$\left. + 2u(B - R - t + c) + 2d\left(B - \frac{t}{2}\right) \right] \frac{t}{A_e}$$

$$= 1.265 \text{ in.}$$

Since the flanges are fairly stocky when treated as webs in bending, they will be fully effective and their effective widths in bending do not need to be calculated.

$$I_{ye} = \left[\left(\frac{t}{2}\right)^2 b_w + 2w\left(\frac{B}{2}\right)^2 + 2u(t+R-c)^2 \right.$$

$$\left. + 2u(B-R-t+c)^2 + 2d\left(B-\frac{t}{2}\right)^2 \right] t$$

$$+ 2t\frac{w^3}{12} - A_e x_{ce}^2 = 0.693 \text{ in.}^4$$

$$S_{cy} = \frac{I_{ye}}{x_{ce}} = 0.548 \text{ in.}^3$$

$$M_{ny} = S_{cy}F_c = 24.65 \text{ kip-in.} \qquad \text{(Eq. C3.1.2-1)}$$

The moment M_{ny} may be greater than the nominal section flexural strength M_{no} since it is based on the effective section modulus for the compression flange at a stress F_c as stated in Section C3.1.2. The nominal section flexural strength for the tension flange at yield also needs to be calculated.

Assume a stress in the compression flange which will produce yield in the tension flange and iterate until the tension flange is at yield as shown in Figure 8.8. Assume $f = 30.8\,\text{ksi}$.

$$k = 4.0$$

$$\lambda = \frac{1.052}{\sqrt{k}}\left(\frac{h}{t}\right)\sqrt{\frac{f}{E}} = 1.028 \qquad \text{(Eq. B2.1-4)}$$

$$\rho = \frac{1 - 0.22/\lambda}{\lambda} = 0.764 \qquad \text{(Eq. B2.1-3)}$$

Since $\lambda > 0.673$, then

$$b_w = \rho h = 2.775 \text{ in.} \tag{Eq. B2.1-2}$$

$$A_e = t(2w + b_w + 4u + 2d) = 0.593 \text{ in.}^2$$

$$x_{ce} = \left[\frac{t}{2} b_w + 2w\left(\frac{B}{2}\right) + 2u(t + R - c) \right.$$

$$\left. + 2u(B - R - t + c) + 2d\left(B - \frac{t}{2}\right) \right] \frac{t}{A_e}$$

$$= 1.218 \text{ in.}$$

$$f_t = f \frac{B - x_{ce}}{x_{ce}} = 45.0 \text{ ksi}$$

Hence, tension flange is at yield and compression flange is at a stress lower than yield.

Since the flanges are fairly stocky when treated as webs in bending, they will be fully effective and their effective widths in bending do not need to be calculated.

$$I_{ye} = \left[\left(\frac{t}{2}\right)^2 b_w + 2w\left(\frac{B}{2}\right)^2 + 2u(t + R - c)^2 \right.$$

$$\left. + 2u(B - R - t + c)^2 + 2d\left(B - \frac{t}{2}\right)^2 \right] t$$

$$+ 2t \frac{w^3}{12} - A_e x_{ce}^2 = 0.725 \text{ in.}^4$$

$$S_{ey} = \frac{I_{ye}}{B - x_{ce}} = 0.407 \text{ in.}^3$$

$$M_{no} = S_{ey} F_y = 18.32 \text{ kip-in.} \tag{Eq. C3.1.1-1}$$

This value of M_{no} is less than $M_{ny} = 24.65$ kip-in. computed above since the tension flange is at yield. Hence,

$$M_{ny} = M_{no} = 18.32 \text{ kip-in.}$$

D. Combined Axial Compressive Load and Bending
    Section C5.2 For uniform moment

$$C_{my} = C_m = 1.0$$

Centroid of effective section under axial force alone from Section 7.6.3D:

$$x_{cea} = 0.958 \text{ in.}$$

Use P_{no}, P_n from Example 7.6.3: $P_{no} = 19.74\,\text{kips}$, $P_n = 15.75\,\text{kips}$.

Clause C5.2.1 ASD method

$$\Omega_c = 1.80 \qquad \Omega_b = 1.67$$

Try

$$P = 4.8 \text{ kips}$$

$$M_y = Px_{cea} = 4.60 \text{ kip-in.}$$

$$A = 0.645 \text{ in.}^2 \qquad\qquad \text{(Example 7.6.3)}$$

$$\sigma_{ey} = 223.6 \text{ ksi}$$

$$P_{Ey} = A\sigma_{ey} = 144.1 \text{ kips}$$

$$\alpha_y = 1 - \frac{\Omega_b P}{P_{Ey}} = 0.940 \qquad\qquad \text{(Eq. C5.2.1-5)}$$

$$\frac{\Omega_c P}{P_n} + \frac{\Omega_b C_{my} M_y}{M_{ny}\alpha_y} = \frac{1.80 \times 4.8}{15.75} + \frac{1.67 \times 1 \times 4.60}{18.32 \times 0.940} = 1.0$$

$$\text{(Eq. C5.2.1-1)}$$

$$\frac{\Omega_c P}{P_{no}} + \frac{\Omega_b M_y}{M_{ny}} = \frac{1.80 \times 4.8}{19.74} + \frac{1.67 \times 4.60}{18.32} = 0.857$$

$$\text{(Eq. C5.2.1-2)}$$

Hence, the nominal axial compressive load applied eccentrically at web is

$$P = 4.8 \text{ kips}$$

Section C5.2.2 LRFD method

$$\phi_c = 0.85 \qquad \phi_b = 0.90$$

Try

$$P_u = 7.35 \text{ kips}$$

$$M_{uy} = P_u x_{cea} = 7.04 \text{ kip-in.}$$

$$P_{Ey} = A\sigma_{ey} = 144.1 \text{ kips}$$

$$\alpha_y = 1 - \frac{P_u}{P_{Ey}} = 0.949 \qquad \text{(Eq. C5.2.2-5)}$$

$$\frac{P_u}{\phi_c P_n} + \frac{C_{my}M_{uy}}{\phi_b M_{ny}\alpha_y} = \frac{7.35}{0.85 \times 15.75} + \frac{7.04}{0.90 \times 18.32 \times 0.949}$$

$$= 1.0 \qquad \text{(Eq. C5.2.2-1)}$$

$$\frac{P_u}{\phi_c P_{no}} + \frac{M_{uy}}{\phi_b M_{ny}} = \frac{7.35}{0.85 \times 19.74} + \frac{7.04}{0.90 \times 18.32}$$

$$= 0.865 < 1.0$$

$$\text{(Eq. C5.2.2-2)}$$

Hence, the required axial strength for a load applied eccentrically at web is

$$P_u = 7.35 \text{ kips}$$

REFERENCES

8.1 Trahair, N. S. and Bradford, M. A., The Behaviour and Design of Steel Structures, 3rd ed., Chapman and Hall, London, 1998.

8.2 Peköz, T. and Winter, G., Torsional-flexural buckling of thin-walled sections under eccentric load, Journal of the Structural Division, ASCE, Vol. 95, No. ST5, May 1969, pp. 1321–1349.

8.3 Jang, D. and Chen, S., Inelastic Torsional-Flexural Buckling of Cold-Formed Open Section under Eccentric Load, Seventh International Specialty Conference on Cold-Formed Steel Structures, St Louis, MO, November, 1984.

8.4 Rack Manufacturers Institute, Specification for the Design, Testing and Utilization of Industrial Steel Storage Racks, Materials Handling Institute, Charlotte, NC, 1997.

8.5 Peköz, T., Unified Approach for the Design of Cold-Formed Steel Structures, Eighth International Specialty Conference on Cold-Formed Steel Structures, St Louis, MO, November, 1986.

8.6 Loh, T. S., Combined Axial Load and Bending in Cold-Formed Steel Members, Ph.D. thesis, Cornell University, February 1985, Report No. 85-3.

9

Connections

9.1 INTRODUCTION TO WELDED CONNECTIONS

Welded connections between thin-walled cold-formed steel sections have become more common in recent years despite the shortage of design guidance for sections of this type. Two publications based on work at Cornell University, New York (Ref. 9.1), and the Institute TNO for Building Materials and Building Structures in Delft, Netherlands (Ref. 9.2), have produced useful test results from which design formulae have been developed. The design rules in the AISI Specification were developed from the Cornell tests. However, the more recent TNO tests add additional information to the original Cornell work, and so the results and design formulae derived in Refs. 9.1 and 9.2 are covered in this chapter even though the AISI Specification is based solely on Ref. 9.1.

Sheet steels are normally welded with conventional equipment and electrodes. However, the design of the connections produced is usually different from that for hot-rolled sections and plate for the following reasons:

(a) Stress-resisting areas are more difficult to define.
(b) Welds such as the arc spot and seam welds in Figures 9.1c and d are made through the welded sheet without any preparation.
(c) Galvanizing and paint are not normally removed prior to welding.
(d) Failure modes are complex and difficult to categorize.

The usual types of fusion welds used to connect cold-formed steel members are shown in Figure 9.1, although groove welds in butt joints may be difficult to produce in thin sheet and are therefore not as common as fillet, spot, seam, and flare groove welds. Arc spot and slot welds are commonly used to attach cold-formed steel decks and panels to their supporting frames. As for conventional structural welding, it is general practice to require that the weld materials be matched at least to the strength level of the weaker member. Design rules for the five weld types in Figures 9.1a–e are given as Sections E2.1–E2.5, respectively, in the AISI Specification.

Failure modes in welded sheet steel are often complicated and involve a combination of basic modes, such as sheet tearing and weld shear, as well as a large amount of out-of-plane distortion of the welded sheet. In general, fillet welds in thin sheet steel are such that the leg length on the sheet edge is equal to the sheet thickness, and the other leg is often two to three times longer. The throat thickness (t_w in Figure 9.2a) is commonly larger than the thickness (t) of the sheet steel, and, hence, ultimate failure is usually found to occur by tearing of the plate adjacent to the weld or along the weld contour. In most cases, yielding is poorly defined and rupture rather than yielding is a more reliable

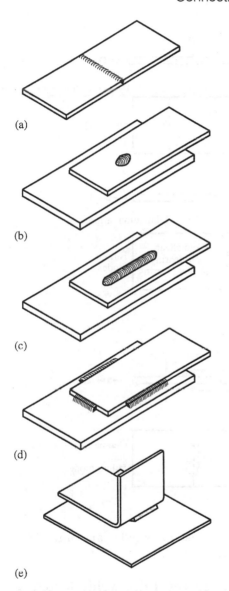

(a)

(b)

(c)

(d)

(e)

FIGURE 9.1   Fusion weld types: (a) groove weld in butt joint;
(b) arc spot weld (puddle weld); (c) arc seam welds; (d) fillet welds;
(e) flare-bevel groove weld.

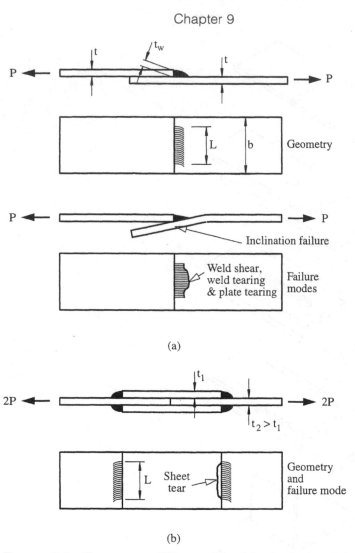

FIGURE 9.2 Transverse fillet welds: (a) single lap joint (TNO tests); (b) double lap joint (Cornell test).

criterion of failure. Hence, for the fillet welds tested at Cornell University and Institute TNO, the design formulae are a function of the tensile strength (F_u) of the sheet material and not of the yield point (F_y). This latter formulation has the added advantage that the yield strength of the cold-formed steel in the heat-affected zone does not play a role in the design and, hence, need not be determined.

As a result of the different welding procedures required for sheet steel, the specification of the American Welding Society for Welding Sheet Steel in Structures (Ref. 9.3) should be closely followed and has been referenced in the AISI Specification. The fact that a welder may have satisfactorily passed a test for structural steel welding does not necessarily mean that he can produce sound welds on sheet steel. The welding positions covered by the Specification are given for each weld type in Table E2 of the Specification.

For the welded connection in which the thickness of the thinnest connected part is greater than 0.18 in., the design rules in the AISC "Load and Resistance Factor Design for Structural Steel Buildings" (Ref. 1.1) should be used. The reason for this is that failure through the weld throat governs for thicker sections rather than sheet tearing adjacent to the weld in thinner sections.

In the case of fillet welds and flare groove welds, failure through the weld throat is checked for plate thicknesses greater than 0.15 in. as specified in Sections E2.4 and Section E2.5.

9.2 FUSION WELDS

9.2.1 Groove Welds in Butt Joints

In the AISI Specification the nominal tensile and compressive strengths and the nominal shear strength are specified for a groove butt weld. The butt joint nominal tensile or compressive strength (P_n) is based on the yield point used in design for the lower-strength base steel and is given by

$$P_n = Lt_eF_y \tag{9.1}$$

where L is the length of the full size of the weld and t_e is the effective throat dimension of the groove weld. A resistance factor of 0.90 is specified for LRFD and a factor of safety of 2.5 is specified for ASD, and they are the same as for a member.

The nominal shear strength (P_n) is the lesser of the shear on the weld metal given by Eq. (9.2) and the shear on the base metal given by Eq. (9.3):

$$P_n = Lt_e(0.6F_{xx}) \tag{9.2}$$

$$P_n = Lt_e\left(\frac{F_y}{\sqrt{3}}\right) \tag{9.3}$$

where F_{xx} is the nominal tensile strength of the groove weld metal. A resistance factor of 0.8 for LRFD is used with Eq. (9.2), and a resistance factor of 0.9 for LRFD is used with Eq. (9.3) since it applies to the base metal. Equation (9.2) applies to the weld metal and therefore has a lower resistance factor than Eq. (9.3). For ASD, a factor of safety of 2.5 is used with both equations.

9.2.2 Fillet Welds Subject to Transverse Loading

The Cornell test data for fillet welds, deposited from covered electrodes, was produced for the type of double lap joints shown in Figure 9.2b. These joints failed by tearing of the connected sheets along or close to the contour of the welds, or by secondary weld shear. Based on these tests, Eq. (9.4) was proposed to predict the connection strength:

$$P_n = tLF_u \tag{9.4}$$

where t is the sheet thickness, L is the length of weld perpendicular to the loading direction, and F_u is the tensile strength of the sheet. The results of these tests are shown in Figure 9.3a for all failure modes where they are compared with the prediction of Eq. (9.4). The values on the abscissa of Figure 9.3a are $2P_n$ since the joints tested were double lap joints. A resistance factor (ϕ) of 0.60 for

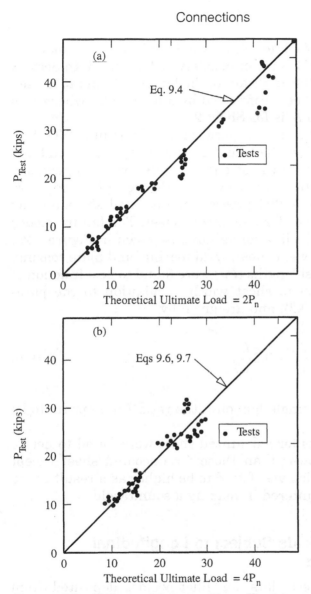

FIGURE 9.3 Fillet weld tests (Cornell): (a) transverse (Figure 9.2b); (b) longitudinal (Figure 9.4b).

LRFD is specified for fillet welds subject to transverse loading. For ASD, a factor of safety of 2.5 is used. In Section E2.4 of the AISI Specification, the lesser of t_1 and t_2 is used to check both sheets connected by a fillet weld, where t_1 is for Sheet 1 and t_2 is for Sheet 2.

A series of tests was performed at Institute TNO (Ref. 9.2) to determine the effect of single lap joints and the welding process on the strength of fillet weld connections. The joints tested are shown in Figure 9.2a and were fabricated by the TIG process for uncoated sheet and by covered electrodes for galvanized sheet. The failure modes observed were inclination failure, as shown in Figure 9.2a, combined with weld shear, weld tearing, and plate tearing. The mean test strengths (P_m) were found to be a function of the weld length to sheet width in addition to the parameters in Eq. (9.2), and are given by

$$P_m = tLF_u\left(1 - 0.3\frac{L}{b}\right) \tag{9.5}$$

Hence, for the single lap joint, the ratio L/b appears to be important.

The results for galvanized sheet were found to not be significantly lower than those for uncoated sheet except that the deviation was found to be higher as a result of the difficulty encountered in making a sound weld.

9.2.3 Fillet Welds Subject to Longitudinal Loading

The Cornell test data for fillet welds, deposited from covered electrodes, was produced for the type of double lap joints shown in Figure 9.4b. These joints failed by tensile tearing across the connected sheet or by weld shear or tearing along the sheet parallel to the contour of the weld, as shown in Figure 9.4b. Based on these tests, the

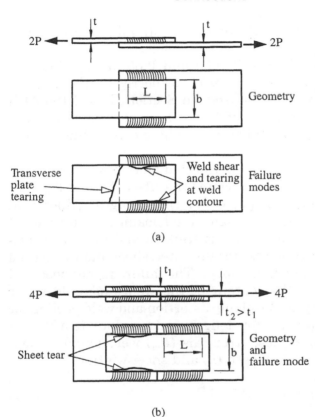

(a)

(b)

FIGURE 9.4   Fillet welds subject to longitudinal loading: (a) single lap joint (TNO tests); (b) double lap joint (Cornell tests).

following equations were found to predict the connection strength:

$$P_n = tL\left(1 - 0.01\frac{L}{t}\right)F_u \qquad \text{for } \frac{L}{t} < 25 \qquad (9.6)$$

$$P_n = 0.75tLF_u \qquad \text{for } \frac{L}{t} \geqslant 25 \qquad (9.7)$$

P_n is the strength of a single weld. The results of all tests are shown in Figure 9.3b compared with the predictions of Eqs. (9.6) and (9.7) where the least value has been used. The values on the abscissa of Figure 9.3b are $4P_n$ since the joints tested were double lap joints with fillet welds on each side of each sheet.

Resistance factors of 0.60 and 0.55 are specified for LRFD for Eqs. (9.6) and (9.7), respectively. For ASD, a factor of safety of 2.5 is used. In Section E2.4 of the AISI Specification, the lesser of t_1 and t_2 is used to check both sheets connected by a fillet weld, where t_1 is for Sheet 1 and t_2 is for Sheet 2.

A series of tests was performed more recently at Institute TNO (Ref. 9.2) to determine the effect of single lap joint geometry and welding process on the strength of fillet weld connections subject to longitudinal loading. The types of joints tested are shown in Figure 9.4a and were manufactured by the TIG process for uncoated sheet and by covered electrodes for galvanized sheet. The failure modes observed were tearing at the contour of the weld and weld shear accompanied by out-of-plane distortion and weld peeling for short-length welds. For longer welds, plate tearing occurred. The mean test strengths (P_m) were found to be a function of the weld length (L) and sheet width (b):

$$P_m = 2tL\left(0.95 - 0.45\frac{L}{b}\right)F_u \tag{9.8}$$

$$P_m = 0.95tbF_u \tag{9.9}$$

The lesser of the values of Eq. (9.8) for weld failure and Eq. (9.9) for plate failure should be used. As for the transverse fillet welds, there was no significant difference between the values for the uncoated and galvanized sheets.

9.2.4 Combined Longitudinal and Transverse Fillet Welds

Tests were performed at Institute TNO (Ref. 9.2) to ascertain whether combined longitudinal and transverse fillet welds interacted adversely or beneficially. The test series showed that similar failure modes to those for longitudinal and transverse fillet welds were observed. Also, the defor-

mation capacity of the individual welds allowed full coop-
eration so that the strengths of the transverse and longitudinal welds can be simply added. In fact, the combined
welds were better than the sum of the two, but the additional benefits have not been quantified.

9.2.5 Flare Groove Welds

Flare welds are of two common types. These are flare-bevel
welds, as shown in Figure 9.1e and in cross section in
Figure 9.5a, and flare V-welds, as shown in cross section
in Figure 9.5b.

As for fillet welds, the flare welds may be loaded
transversely or longitudinally, and their modes of failure
are similar to fillet welds described in Sections 9.2.2 and
9.2.3. The primary mode of failure is sheet tearing along
the weld contour.

For transverse flare-bevel groove welds, the nominal
shear strength in Section E2.5 of the AISI Specification is
the same as for a fillet weld subject to transverse loading
[Eq. (9.4)] except that it is factored by $5/6 = 0.833$:

$$P_n = 0.833tLF_u \qquad (9.10)$$

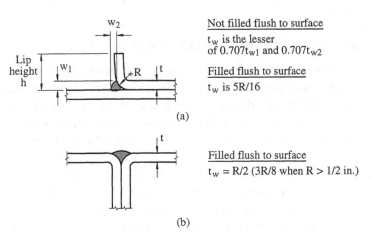

w_2

Not filled flush to surface

t_w is the lesser
of $0.707t_{w1}$ and $0.707t_{w2}$

Lip
height
h

w_1

R

t

Filled flush to surface

t_w is $5R/16$

(a)

t

Filled flush to surface

$t_w = R/2$ (3R/8 when R > 1/2 in.)

(b)

FIGURE 9.5 Flare groove weld cross sections: (a) flare-bevel
groove weld; (b) flare V-groove weld.

The resistance factor to be used in LRFD with Eq. (9.10) is $\phi = 0.55$. For ASD a factor of safety of 2.5 is used.

For flare groove welds subject to longitudinal loading, either on a flare-bevel groove weld as in Figure 9.5a or a flare V-groove weld as in Figure 9.5b, the nominal shear strength depends on the effective throat thickness (t_w) of the flare weld and the lip height (h) in Figure 9.5a. If $t \leqslant t_w < 2t$ or the lip height (h) is less than the weld length (L), the nominal strength is the same for a fillet weld subject to longitudinal loading given by Eq. (9.7), so that

$$P_n = 0.75tLF_u \tag{9.11}$$

If $t_w \geqslant 2t$ and $h \geqslant L$, then

$$P_u = 1.5tLF_u \tag{9.12}$$

The resistance factor to be used in both cases in LRFD is 0.55, and the factor of safety in ASD is 2.5.

In the AISI Specification, Section E2.5 defines t_w for a range of cases including flare groove welds filled flush to surface and flare groove welds not filled flush to surface, as shown in Figure 9.5. For $t > 0.15$ in., failure through the throat thickness t_w is checked based on the weld metal strength of F_{xx}.

9.2.6 Arc Spot Welds (Puddle Welds)

Arc spot welds are for welding sheet steel to thicker supporting members in a flat position. Arc spot welds have three different diameters at different levels of the weld, as shown in Figure 9.6a for a single thickness of attached sheet and in Figure 9.6b for a double thickness of attached sheet. The diameter d is the visible width of the arc spot weld, the diameter d_a is the average diameter at mid-thickness, and the diameter d_e is the effective diameter of the fused area. In general, arc-spot-welded connections are applied in sheet steel with thicknesses from about 0.02 in. to 0.15 in. Weld washers must be used when the thickness of the sheet is less than 0.028 in. In many ways, the design of an arc spot weld is similar to an equivalent bolted connection since the failure modes are similar to those of mechanical connections. The principal modes of failure are shown in Figure 9.7. Only the mode shown in (b), tearing and bearing at weld contour, does not have an exact equivalent in bolted connection design.

The Cornell test data for spot welds is shown in Figure 9.8a. Those joints which failed by weld shear, as shown in Figure 9.7e, were used to determine the effective diameter (d_e) of the fused area:

$$d_e = 0.70d - 1.5t \leqslant 0.55d \tag{9.13}$$

The minimum allowable value of d_e is $\frac{3}{8}$ in.

Weld shear failure loads can be successfully predicted by

$$V_w = \frac{\pi}{4}(d_e)^2 0.75F_{xx} \tag{9.14}$$

where F_{xx} is the filler metal strength for the AWS Electrode Classification. A resistance factor of 0.60 for LRFD is specified in the AISI Specification for use with Eq. (9.14). For ASD a factor of safety of 2.5 is used.

For plate tearing around the weld, as shown in Figure 9.7b, the following formulae for the nominal shear strength of an arc spot weld have been developed:

$$P_n = 2.2td_aF_u \quad \text{for } \frac{d_a}{t} \leqslant 0.815\sqrt{\frac{E}{F_u}} \tag{9.15}$$

$$P_n = 1.4td_aF_u \quad \text{for } \frac{d_a}{t} \geqslant 1.397\sqrt{\frac{E}{F_u}} \tag{9.16}$$

and

$$P_n = 0.280\left(1 + \frac{5.59\sqrt{E/F_u}}{d_a/t}\right)td_aF_u$$

$$\text{in the transition region} \tag{9.17}$$

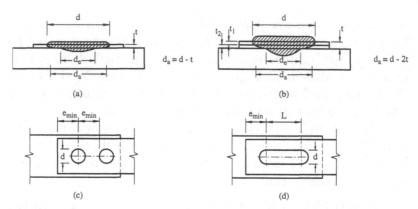

FIGURE 9.6 Arc spot and arc seam weld geometry: (a) single thickness of sheet; (b) double thickness of sheet; (c) minimum edge distance (arc spot welds); (d) geometry and minimum edge distance (arc seam welds).

where d_a is the average diameter of the arc spot weld, F_u is the tensile strength of the sheet material, and t is the total combined sheet thickness of steels involved in shear transfer above the plane of maximum shear transfer. The value of d_a is $d - t$ for a single sheet and $d - 2t$ for multiple sheets.

The results of all of the tests are shown in Figure 9.8a compared with the predictions of Eqs. (9.13)–(9.19) where the least value has been used. The values on the abscissa of Figure 9.8a are $2P_n$ since the joints tested were double lap joints. The variability of the test results in Figure 9.8a occurred because all of the field-welded specimens were poorly made. Resistance factors in LRFD of 0.60, 0.50, 0.50 apply to Eqs. (9.15)–(9.17), respectively, as specified in Section E2.2.1 of the AISI Specification. For ASD a factor of safety of 2.5 is used throughout.

The failure mode of net section failure shown in Figure 9.7d must be checked separately by using the net section capacity of a tension member as given by Section C2 of the AISI Specification. For the purpose of this calculation, the net section is the full section width (b) of the plate.

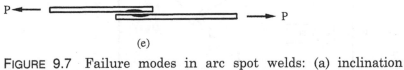

(a)

(b)

(c)

(d)

(e)

FIGURE 9.7 Failure modes in arc spot welds: (a) inclination failure; (b) tearing and bearing at weld contour; (c) edge failure; (d) net section failure; (e) weld shear failure.

The failure mode of edge failure shown in Figure 9.7c is limited in Section E2.2.1 of the AISI Specification by limiting the edge distance. The minimum distance e_{min} from the centerline of a weld to the nearest edge of an adjacent weld or the end of the connected point toward which the force is directed is

$$e_{min} = \frac{P_u}{\phi F_u t} \qquad \text{for LRFD} \qquad (9.18)$$

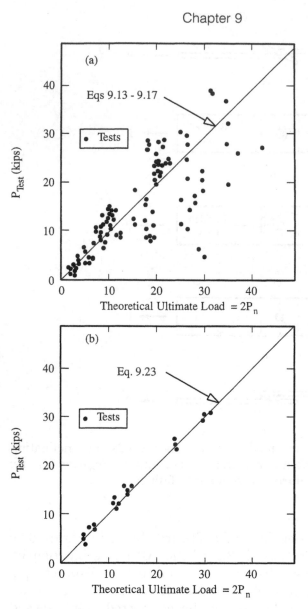

FIGURE 9.8   Arc spot and arc seam weld tests (Cornell): (a) arc spot welds; (b) arc seam welds.

where P_u is the required strength (factored) transmitted by the weld.

$$e_{\min} = \frac{P\Omega}{F_u t} \quad \text{for ASD} \tag{9.19}$$

where P is the required strength (unfactored) transmitted by the weld. In Eqs. (9.18) and (9.19), t is the thickness of the thinnest connected sheet. The values of resistance factor (ϕ) and factor of safety (Ω) depend on the ratio of F_u/F_{sy}: when it is less than 1.08, ϕ and Ω are 0.70 and 2.0, respectively; otherwise they are 0.60 and 2.22, respectively.

Arc spot welds in tension are specified in Section E2.2.2 of the AISI Specification. Either weld failure or sheet tear around the weld can occur. For weld failure, the nominal tensile strength (P_n) of each concentrically loaded arc spot weld is given by

$$P_n = \frac{\pi}{4} d_e^2 F_{xx} \tag{9.20}$$

such that F_{xx} is the filler metal strength, which must be greater than 60 ksi and F_u. For sheet tear the nominal tensile strength (P_n) of each concentrically loaded arc spot weld is given by

$$P_n = \left[6.59 - \frac{3150 F_u}{E}\right] t d_a F_u \leqslant 1.46 t d a F_u$$
$$\text{for } F_u < 0.00187E \tag{9.21}$$

$$P_n = 0.70 t d_a F_u \quad \text{for } F_u \geqslant 0.00187E \tag{9.22}$$

where $F_u \leqslant 82$ ksi. The values of resistance factor for LRFD and safety factor for ASD are 0.60 and 2.50, respectively. Further, if eccentric loading can be applied to the arc spot weld, then it is prudent to reduce the tensile capacity of an arc spot weld to 50% of the values given by Eqs. (9.20) (9.22).

9.2.7 Arc Seam Welds

Arc seam welds commonly find application in the narrow troughs of cold-formed steel decks and panels. The geometry of an arc seam weld is shown in Figure 9.6d. The observed behavior of the arc seam welds tested at Cornell was similar to the arc spot welds, although no simple shear failures were observed. The major mode of failure was tensile tearing of the sheets along the forward edge of the weld contour plus shearing of the sheets along the sides of the weld. The full set of test results for the arc welds is shown in Figure 9.8b. In general, there was considerably less variability than for the arc spot welds shown in Figure 9.8a.

Based on the Cornell tests, the following formula has been developed for arc seam welds:

$$P_n = 2.5tF_u(0.25L + 0.96d_a) \tag{9.23}$$

where d_a is the width of the arc seam weld as discussed in Section 9.2.6. The test results plotted in Figure 9.8b are compared with the prediction of Eq. (9.23). The values on the abscissa of Figure 9.8b are $2P_n$ since the joints tested were double lap joints. A resistance factor of 0.60 is specified for LRFD for use with Eq. (9.23). The factor of safety for ASD is 2.50.

In addition, the AISI Specification includes a formula to check the shear capacity of the weld:

$$P_n = \left[\frac{\pi d_e^2}{4} + Ld_e\right]0.75F_{xx} \tag{9.24}$$

where F_{xx} is the filler metal strength in the AWS electrode classification.

The minimum edge distance (e_{min}) shown in Figure 9.6d is the same as specified for arc spot welds described by Eqs. (9.18) and (9.19).

9.3 RESISTANCE WELDS

Resistance welds are a type of arc spot weld and, as such, may fail in shear. The nominal shear capacities (P_n) of resistance spot welds specified in Table E2.6 of the AISI Specification were simply taken from Ref. 9.4. A resistance factor of 0.65 is specified for LRFD in Section E2.6 of the AISI Specification. Welding shall be performed in accordance with AWS C1.3-70 (Ref. 9.4) and AWS C1.1-66 (Ref. 9.5) and the AISI Specification, and a factor of safety of 2.5 is specified for ASD.

By comparison, the approach developed at the Institute TNO (Ref. 9.2) uses the same design formulae for resistance welds as for arc spot welds but with a different formula for the effective weld diameter from that for arc spot welds.

9.4 INTRODUCTION TO BOLTED CONNECTIONS

Bolted connections between cold-formed steel sections require design formulae different from those of hot-rolled construction, as a result of the smaller ratio of sheet thickness to bolt diameter in cold-formed design. The design provisions for bolted connections in the AISI Specification are based mainly on Ref. 9.6.

The AISI Specification allows use of bolts, nuts, and washers to the following ASTM specifications.

> ASTM A194/A194M   Carbon and Alloy Steel Nuts for Bolts for High- Pressure and High-Temperature Service
>
> ASTM A307 (Type A)   Carbon Steel Bolts and Studs, 60,000 PSI Tensile Strength
>
> ASTM A325   Structural Bolts, Steel, Heat Treated, 120/150 ksi Minimum Tensile Strength
>
> ASTM A325M   High Strength Bolts for Structural Steel Joints [metric]

ASTM A354 (Grade BD) Quenched and Tempered Alloy Steel Bolts, Studs, and Other Externally Threaded Fasteners (for diameter of bolt smaller than $\frac{1}{2}$ in.)

ASTM A449 Quenched and Tempered Steel Bolts and Studs (for diameter of bolt smaller than $\frac{1}{2}$ in.)

ASTM A490 Heat-Treated Steel Structural Bolts, 150 ksi Minimum Tensile Strength

ASTM A490M High Strength Steel Bolts, Classes 10.9 and 10.9.3, for Structural Steel Joints [metric]

ASTM A563 Carbon and Alloy Steel Nuts

ASTM A563M Carbon and Alloy Steel Nuts [metric]

ASTM F436 Hardened Steel Washers

ASTM F436M Hardened Steel Washers [metric]

ASTM F844 Washers, Steel, Plain (Flat), Unhardened for General Use

ASTM F959 Compressible Washer-Type Direct Tension Indicators for Use with Structural Fasteners

ASTM F959M Compressible Washer-Type Direct Tension Indicators for Use with Structural Fasteners [metric]

Bolts manufactured according to ASTM A307 have a nominal tensile strength of 60 ksi. The standard size range defined in A307 is from $\frac{1}{4}$ to $\frac{3}{4}$ in. in $\frac{1}{16}$-in. increments. Bolts manufactured according to ASTM A325 and ASTM A449 have a nominal tensile strength of 120 ksi and a nominal 0.2% proof stress of 92 ksi. The standard size range defined in ASTM A325 is $\frac{1}{2}$ to $\frac{3}{4}$ in. in $\frac{1}{16}$-in. increments; in ASTM A449 the range is $\frac{1}{4}$ in. to $\frac{7}{16}$ in. in $\frac{1}{16}$-in. increments. The design rules in the AISI Specification only apply when the thickness of a connected part is less than $\frac{3}{16}$ in. For connected parts $\frac{3}{16}$ in. or greater, the AISC Specification (Ref. 1.1) should be used.

A selection of bolted connections, where the bolts are principally in shear, is shown in Figure 9.9. As shown in

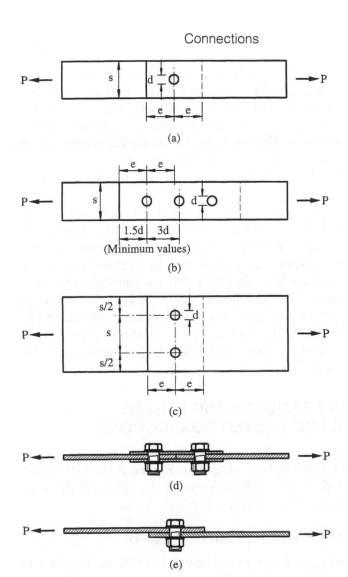

FIGURE 9.9 Bolted connection geometry: (a) single bolt ($r = 1$); (b) three bolts in line of force ($r = \frac{1}{3}$); (c) two bolts across line of force ($r = 1$); (d) double shear (with washers); (e) single shear (with washers).

this figure, the bolts may be located in line with the line of action of the force (Figure 9.9b) or perpendicular to the line of action of the force (Figure 9.9c) or both simultaneously. The connections may be in double shear with a cover plate on each side, as in Figure 9.9d, or in single shear, as in Figure 9.9e. In addition, each bolt may contain washers under the head and nut, under the nut alone, or under neither.

The symbols e for edge distance or distance to an adjacent bolt hole, s for bolt spacing perpendicular to the line of stress, and d for bolt diameter are shown for the various bolted connection geometries in Figures 9.9a–c. In Table E3, the AISI Specification defines the diameter of a standard hole (d_h) as $\frac{1}{16}$ in. larger than the bolt diameter (d) for bolt diameters $\frac{1}{2}$ in. and larger and $\frac{1}{32}$ in. larger than the bolt diameter for bolts less than $\frac{1}{2}$ in. diameter. Oversized holes, short-slotted holes, and long-slotted holes are also defined in Table E3.

9.5 DESIGN FORMULAE AND FAILURE MODES FOR BOLTED CONNECTIONS

The four principal modes of failure for bolted connections of the types in Figure 9.9 are shown in Figure 9.10. They have been classified in Ref. 9.6 as Types I, II, III, and IV. This classification has been followed in this book.

9.5.1 Tearout Failure of Sheet (Type I)

For sheets which have a small distance (e) from the bolt to the edge of the plate or an adjacent hole where the distance is taken in the direction of the line of action of the force, sheet tearing along two lines as shown in Figure 9.10a can occur. Using the tests summarized in Ref. 9.6, the following relationship has been developed to predict tearout failure:

$$\frac{F_{bu}}{F_u} = \frac{e}{d} \tag{9.25}$$

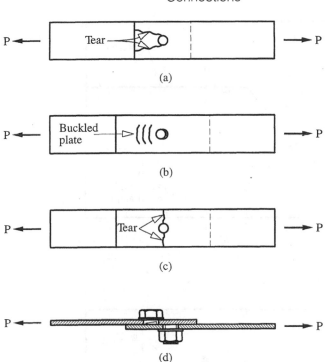

FIGURE 9.10 Failure modes of bolted connections: (a) tearout failure of sheet (Type I); (b) bearing failure of sheet material (Type II); (c) tension failure of net section (Type III); (d) shear failure of bolt (Type IV).

where F_{bu} is the bearing stress between the bolt and sheet at tearout failure. Equation (9.25) is compared with the tests on bolted connections in single shear (with and without washers) in Figure 9.11. Additional graphs of test results for double shear are given in Ref. 9.6. In all cases, Eq. (9.25) adequately predicts failure for low values of the ratio e/d.

Letting $F_{bu} = P_n/dt$ in Eq.(9.25) results in

$$P_n = teF_u \qquad (9.26)$$

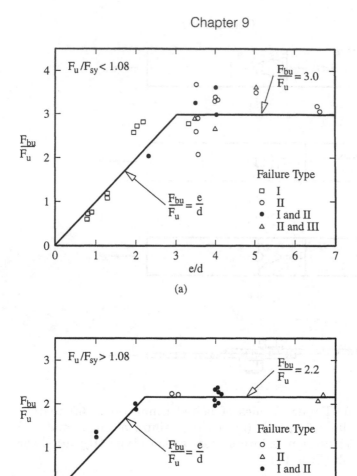

FIGURE 9.11 Bolted connection tests (tearout and bearing failures): (a) single shear connections with washers; (b) single shear connections without washers.

The resistance factor (ϕ) (LRFD) or factor of safety (Ω) (ASD) for use with P_n specified in Section E3.1 of the AISI Specification depends on the ratio of tensile stress to yield stress of the steel, so that

$$\phi = 0.70 \qquad \text{for } \frac{F_u}{F_{sy}} \geq 1.08 \tag{9.27}$$

$$\Omega = 2.0$$

$$\phi = 0.60 \qquad \text{for } \frac{F_u}{F_{sy}} < 1.08 \tag{9.28}$$

$$\Omega = 2.22$$

In addition, the AISI Specification specifies a minimum distance between bolt holes of not less than $3d$ and a minimum distance from the center of any standard hole to the end boundary of a connecting member as $1.5d$.

9.5.2 Bearing Failure of Sheet (Type II)

For bolted connections where the edge distance is sufficiently large, the bearing stress at failure (F_{bu}) is no longer a function of edge distance (e) and bearing failure of the sheet (Figure 9.10b) may occur. The test results shown in Figure 9.11 demonstrate the independence of F_{bu} from e for larger values of e. For the cases of single shear connections with and without washers, the bearing stress at ultimate is shown to be a function of whether washers are installed and the ratio of F_u/F_{sy}. Additional graphs in Ref. 9.6 demonstrate a similar dependence for double-shear connections with and without washers. Hence, the inclusion of washers under both bolt head and nut is important in the design of bolted connections in cold-formed steel, probably as a result of the low sheet thicknesses in cold-formed design.

From Figure 9.11a the nominal bearing capacity for single shear with washers and $F_u/F_{sy} < 1.08$ is

$$P_n = 3.00F_u dt \tag{9.29}$$

The resistance factor for use in LRFD with Eq. (9.29) is specified in Table E3.3-1 of the AISI Specification as 0.60, and the factor of safety for use in ASD is 2.22. Other cases, based on the tests summarized in Ref. 9.6, are set out in Tables E3.3-1 and E3.3-2 of the AISI Specification.

Recent research has been performed at the University of Sydney by Rogers and Hancock (Ref. 9.7) on bolted connections of thin G550 (80 ksi) and G300 (44 ksi) sheet steels in 0.42-mm (0.016 in.) and 0.60-mm (0.024 in.) base metal thickness. The test results indicate that the connection provisions described above cannot be used to accurately predict the failure mode of bolted connections from thin G550 and G300 steels. Furthermore, the design rules cannot be used to accurately determine the bearing resistance of bolted test specimens based on a failure criterion for predicted loads. It is necessary to incorporate a variable resistance equation which is dependent on the thickness of the connected material similar to CAN/CSA-S136-94 (Ref. 1.6). In addition, the ultimate bearing stress to ultimate material strength ratios show that a bearing equation coefficient of less than 2.0 in Eq. (9.29) may be appropriate for G550 sheet steels where $d/t \geq 15$.

9.5.3 Net Section Tension Failure (Type III)

If the stresses in the net section are too high, tension failure of the net section, as shown in Figure 9.10c, may occur before tearout or bearing failure of the sheet. The stresses in the net section are dependent upon stress concentrations adjacent to the bolt holes and hence are a function of the bolt spacing (s) and the number of bolts, in addition to the net area and the load in the section. Consequently, empirical formulae have been developed to relate the stress in the net section to the above parameters. For the case of bolts in single shear, the test results summarized in Ref. 9.6 are shown in Figure 9.12a for connections with washers, and in Figure 9.12b for connec-

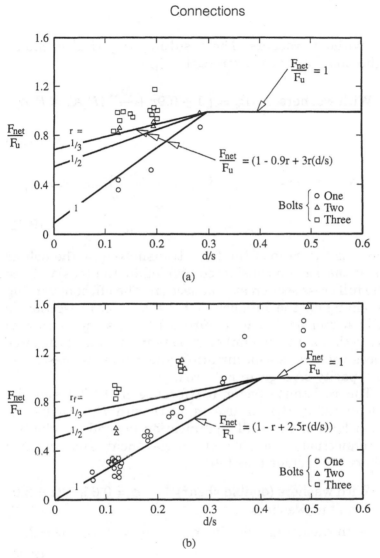

FIGURE 9.12 Bolted connection tests (net section failures): (a) single shear with washers; (b) single shear without washers.

tions without washers. The resulting empirical formulae for the nominal tension strength (P_n) are

With washers $\quad P_n = \left(1 - 0.9r + \dfrac{3rd}{s}\right)F_u A_n \leqslant F_u A_n$

$$(9.30)$$

No washers $\quad P_n = \left(1 - r + \dfrac{2.5rd}{s}\right)F_u A_n \leqslant F_u A_n$

$$(9.31)$$

where r is the ratio of the force transmitted by the bolt or bolts at the section considered divided by the tension force in the full cross section at that section. The effect of varying r by changing the number of bolts at a cross section is clearly demonstrated in Figure 9.12. It is interesting to observe that a larger number of bolts in the line of the force reduces the stress concentration and hence increases the load capacity on a given net area.

The resistance factor (ϕ) for use in LRFD and the factor of safety (Ω) for use in ASD with P_n specified in Section E3.2 of the AISI Specification depend on whether the connection is double or single shear and whether washers are included, as follows:

With washers (double shear) $\qquad \phi = 0.65, \quad \Omega = 2.0$
    and no washers $\hfill (9.32)$

With washer (single shear) $\qquad \phi = 0.55, \quad \Omega = 2.22$

$$(9.33)$$

Based on Ref. 9.7, the net section failure of 0.42-mm (0.016 in.) and 0.60-mm (0.024 in.) G550 (80 ksi) and G300 (44 ksi) sheet steels at connections can be accurately and reliably predicted with the use of the stress reduction factor based on the configuration of bolts and specimen width as given by Eqs. (9.30) and (9.31), and without the use of the

75% stress reduction factor as specified in Section A3.3.2 of the AISI Specification for Grade 80 and Grade E steels.

9.5.4 Shear Failure of Bolt (Type IV)

The nominal shear strength (P_n) of a bolt specified in Section E3.4 of the AISI Specification is given by

$$P_n = A_b F_{nv} \tag{9.34}$$

where F_{nv} is given in Table E3.4-1 of the AISI Specification
A_b is the gross cross-sectional area of the bolt

The value of F_{nv} depends upon whether threads are excluded from the shear plane. The resistance factor for LRFD specified in Table E3.4-1 of the AISI Specification is 0.65, and the factor of safety for ASD is 2.4.

The nominal tension strength (P_n) of a bolt specified in Section E3.4 of the AISI Specification is

$$P_n = A_b F_{nt} \tag{9.35}$$

where F_{nt} is given in Table E3.4-1 of the AISI Specification
A_b is the gross cross-sectional area of the bolt

The resistance factor for LRFD specified in Table E3.4-1 of the AISI Specification is 0.75, and the factor of safety for ASD is generally 2.0, although 2.25 is used for ASTM A307 bolts.

When a combination of shear and tension occurs, the value of F_{nt} in Eq. (9.35) is changed to F'_{nt}. Value of F'_{nt} are given in Tables E3.4-3 and E3.4-5 for LRFD and E3.4-2 and E3.4-4 for ASD. They take the form

$$F'_{nt} = 101 - 2.4 f_v \leqslant 81 \text{ ksi} \tag{9.36}$$

which is the equation for an ASTM A449 bolt designed by LRFD with threads not excluded from the shear plane, with f_v equal to the shear stress.

9.6 SCREW FASTENERS

Screw fasteners are used very frequently in cold-formed steel structures, normally to fasten sheeting to thicker material such as purlins, or to fasten sheets together. A considerable amount of information is available in the international literature and has been used to develop Section E4, Screw Connections of the AISI Specification.

The ECCS document European Recommendations for the Design of Light Gauge Steel Members (Ref. 9.8) was the first to provide detailed guidance on the design of screw fasteners and blind rivets. The background research to this document is set out in Ref. 9.9. Further developments of the European design rules have been performed to produce the design requirements in Eurocode 3 Part 1.3 (Ref. 1.7). In order to apply the methods developed for screwed connections in Ref. 9.8 to U.S. practice, Peköz (Ref 9.10) modified the formulae and recalibrated the results to develop specific design rules for use in the American Iron and Steel Institute Specification. The results are based on a study of screw diameters ranging from 0.11 in. to 0.28 in. In the study, if materials of different thickness were connected, the thinner material was assumed to be in contact with the head of the screw. Two different sets of formulae are provided. These are

1. Design for shear
2. Design for tension
    2.1 Pullout failure
    2.2 Pullover failure

Screw connections *loaded in shear* can fail in one mode or in a combination of modes, including screw shear, edge tearing, tilting, and subsequent pullout of the screw and bearing of the parent plates. The formulae developed only apply where the edge distance (e_1), which is the distance from the center of the screw to the free edge in the direction of loading, or the pitch, p_1 or p_2, which is the center-to-

Connections

center distance of adjacent screws, as shown in Figure
9.13c, is not less than three times the diameter of the
screw (d). However, the other edge distance in the direction
perpendicular to the load (e_2) can be $1.5d$. The formulae for
the nominal shear strength given in Section E4.3-1 of the
AISI Specification are as follows:

(a) When $t_2 \leqslant t_1$, use the smallest of

$$P_{ns} = 4.2F_{u2}\sqrt{t_2^3 d} \tag{9.37}$$

$$P_{ns} = 2.7t_1 dF_{u1} \tag{9.38}$$

$$P_{ns} = 2.7t_2 dF_{u2} \tag{9.39}$$

(b) When $t_2 \geqslant 2.5t_1$, use the smaller of

$$P_{ns} = 2.7t_1 dF_{u1} \tag{9.40}$$

$$P_{ns} = 2.7t_2 dF_{u2} \tag{9.41}$$

(c) When $2.5t_1 > t_2 > t_1$, use linear interpolation
between (a) and (b), where

t_1 = thickness of member in contact with screw head
t_2 = thickness of member *not* in contact with screw head
F_{u1} = tensile strength of steel sheet with thickness t_1
F_{u2} = tensile strength of steel sheet with thickness t_2
d = nominal screw diameter (see Figure 9.13b)

The resistance factor for LRFD is 0.50, and the factor of
safety for ASD is 3.0. Section E4.3.2 requires that the
nominal shear strength of the screw be at least $1.25P_{ns}$.
The shear strength of the screw is determined by test,
according to Section F1.

Failure of the net section based on the net cross-
sectional area of the sheet material should also be checked
in the direction of loading according to Sections E3.2 and
C2 of the AISI Specification.

Screw connections *loaded in tension* can fail by the
screw pulling out of the plate (pullout), by the sheeting
pulling over the screw head and washer if any is present

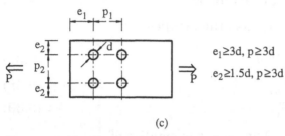

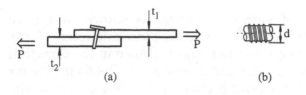

(a)   (b)

(c)

FIGURE 9.13 Screws in shear: (a) thickness; (b) nominal screw diameter (d); (c) minimum edge distances and pitches.

(pullover), or by tensile failure of the screw. The formulae for the nominal pullout capacity (P_{not}) and pullover capacity (P_{nov}) are as follows:

$$P_{not} = 0.85t_c dF_{u2} \tag{9.42}$$

where t_c is the lesser of the depth of penetration and the thickness (t_2) and t_2, F_{u2} are the thickness and tensile strength, respectively, of the sheet not in contact with the head, as shown in Figure 9.14, and

$$P_{nov} = 1.5t_1 d_w F_{u1} \tag{9.43}$$

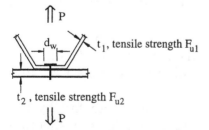

FIGURE 9.14 Screws in tension.

where d_w is the larger of the screw head diameter or the washer diameter but not greater than $\frac{1}{2}$ in. and t_1, F_{u1} are the thickness and tensile strength, respectively, of the sheet adjacent to the head of the screw, as shown in Figure 9.14.

Section E4.4.3 requires the nominal tension strength of the screw to be at least 1.25 times the lesser of P_{not} and P_{nov}. The nominal tension strength of the screw is determined by test according to Section F1.

9.7 RUPTURE

A mode of failure can occur at bolted connections, where tearing failure can occur along the perimeter of holes as shown in Figures 9.15 and 9.16. In some configurations, such as the coped beam with both flanges coped as in Figure 9.15a, the net area of the web in shear resists rupture, and Section E5.1 of the AISI Specification should be used. In other cases, such as the coped beam with one flange coped as shown in Figure 9.15b, block shear rupture occurs where a block is pulled out of the end of the section with a combination of shear failure on one plane and tensile failure on the other plane. For the case of tension loading, block shear rupture planes are shown in Figure 9.16.

Connection tests were performed by Birkemoe and Gilmor (Ref. 9.11) on coped beams to derive formulae for shear rupture and block shear rupture. These were adopted in the American Institute of Steel Construction

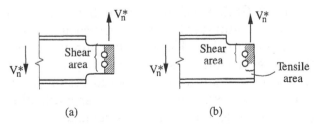

(a)   (b)

FIGURE 9.15   Rupture of coped beam: (a) shear alone; (b) shear and tension.

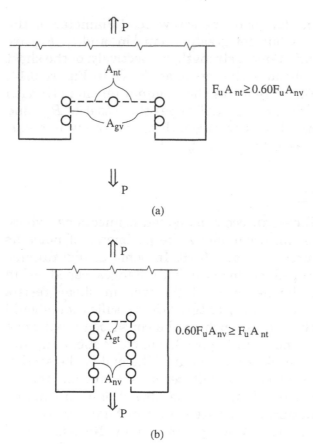

FIGURE 9.16   Block shear rupture: (a) small shear force and large tension force; (b) large shear force and small tension force.

Specification in 1978 (Ref. 9.12) and have been used in various forms in other standards and specifications ever since. They were added to the AISI Specification in 1999.

Test results suggest that it is conservative to predict the block shear strength by adding the yield strength on one plane to the rupture strength on the perpendicular plane. Hence, two possible block shear strengths can be calculated. These are rupture strength on the net tensile area with shear yield on the gross area as shown in Figure

9.16a, and yield on the gross tensile area with shear rupture on the net shear area as shown in Figure 9.16b. These form the basis of Eqs. (E5.3-1) and (E5.3-2) of the AISI Specification, respectively. The controlling equation is governed by the conditions set out in Section E5.3 and not the smaller value of Eqs. (E5.3-1) and (E5.3-2), which will give an erroneous result. In Case (b) in Figure 9.16b, the total force is controlled mainly by shear so that shear rupture should control. In Case (a) in Figure 9.16a, the total force is controlled mainly by tension so that tension rupture should control.

9.8 EXAMPLES

9.8.1 Welded Connection Design Example

Problem

The 3-in.-wide, 0.1-in.-thick ASTM A570 Grade 45 sheet is to be welded to the 0.2-in. plate shown in Figure 9.17 by

- (a) Longitudinal fillet welds, or
- (b) Combined longitudinal and transverse fillet welds, or
- (c) An arc spot weld, or
- (d) An arc seam weld

Determine the size of each weld to fully develop the design strength of the plate. The example uses the LRFD method. The following section numbers refer to the AISI Specification.

Solution

A. Plate Strength for Full Plate (LRFD Method)

For an ASTM A570 Grade 45 steel

$$F_y = 45 \text{ ksi} \quad \text{and} \quad F_u = 60 \text{ ksi}$$

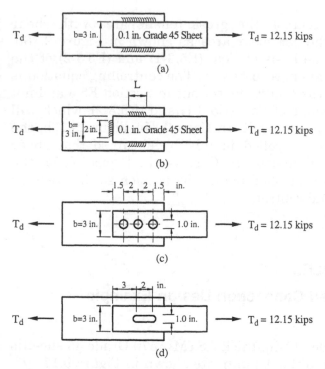

FIGURE 9.17 Weld design example: (a) longitudinal fillet welds; (b) combined longitudinal and transverse fillet welds; (c) arc spot weld; (d) arc seam weld.

Section C2

$$T_n = A_g F_y = bt F_y \qquad \text{(Eq. C2-1)}$$
$$= (3 \times 0.1) \times 45 = 13.5 \text{ kips}$$

Since $\phi_t = 0.90$ for Eq. (C2-1), then

$$\phi_t T_n = 12.15 \text{ kips}$$

$$T_n = A_n F_u = bt F_u \qquad \text{(Eq. C2-2)}$$
$$= (3 \times 0.1) \times 60 = 18 \text{ kips}$$

Since $\phi_t = 0.75$ for Eq. (C2-2), then

$\qquad \phi_t T_n = 0.75 \times 18 = 13.5$ kips

Hence, the design strength of the connection (T_d) is the lesser of $\phi_t T_n = 13.5$ kips and $\phi_t T_n = 12.5$ kips; $T_d = 12.15$ kips.

B.   Longitudinal Fillet Weld Design (LRFD Method)

Section E2.4
Assuming $L/t > 25$,

$\qquad P_n = 0.75tLF_u$ $\qquad\qquad\qquad$ (Eq. E2.4-2)

Now $\phi P_n \leqslant T_d$, where $\phi = 0.55$ for Eq. (E2.4-2). Hence,

$$L = \frac{T_d}{0.55 \times 0.75tF_u}$$

$$= \frac{12.15}{0.55(0.75 \times 0.1 \times 60)}$$

$$= 4.90 \text{ in.}$$

Length of each fillet

$$\frac{L}{2} = \frac{4.90}{2} = 2.45 \text{ in.}$$

Check

$$\frac{L}{t} = \frac{2.45}{0.1} = 24.5 < 25$$

A slightly shorter length could be used according to Eq. (E2.4-1), but a 2.5- in.-weld will be used.

C.   Combined Longitudinal and Transverse Fillet Weld Design (LRFD Method)

First locate transverse fillet weld across 2 in. of end of plate as shown in Figure 9.17b.

Section E2.4

$\qquad (P_n)_{\text{transverse}} - tLF_u = 0.1 \times 2 \times 60 = 12$ kips

$\qquad\qquad\qquad\qquad\qquad\qquad\qquad\qquad$ (Eq. E2.4-3)

Now $\phi = 0.60$ for Eq. (E2.4-3):

$\phi(P_n)_{\text{transverse}} = 7.2$ kips

Hence required

$$(T_d)_{\text{longitudinal}} = T_d - \phi(P_n)_{\text{transverse}}$$

$$= 12.15 - 7.20 = 4.95 \text{ kips}$$

Try $L_l = 0.75$ in. Hence,

$$\frac{L_l}{t} = \frac{0.75}{0.1} = 7.5 < 25$$

$$\phi P_n = 0.60\left(1 - 0.01\frac{L_l}{t}\right)tL_lF_u \qquad \text{(Eq. E2.4-1)}$$

$$= 0.60(1 - 0.01 \times 7.5)0.1 \times 0.75 \times 60 = 2.50 \text{ kips}$$

Hence for a longitudinal fillet weld each side,

$$2\phi P_n = 2 \times 2.50 \text{ kips} = 5.0 \text{ kips} > 4.95 \text{ kips}$$

Hence, use 0.75-in. additional fillet welds on each side in addition to 2-in. transverse fillet weld.

D.   Arc Spot Weld Design (LRFD Method)

Assume three arc spot welds in line as shown in Figure 9.17c, with visible diameter (d) equal to 1.0 in. each with an E60 electrode.

Check Shear Strength of Each Arc Spot Weld, Assuming $F_{xx} = 60$ ksi.

Section E2.2.1

$$d_e = 0.7d - 1.5t \leqslant 0.55d = 0.7 \times 1.0 - 1.5 \times 0.1$$

$$= 0.55 \text{ in.}$$

$$\text{(Eq. E2.2.1-5)}$$

$$\frac{d_e}{d} = 0.55 \leqslant 0.55$$

$$P_n = \frac{\pi}{4}(d_e)^2 0.75F_{xx} \qquad \text{(Eq. E2.2.1-1)}$$

$$= 0.785(0.55)^2 0.75 \times 60 = 10.68 \text{ kips}$$

Now $\phi = 0.60$ for Eq. (2.2.1-1).

$$\phi P_n = 0.60 \times 10.68 = 6.41 \text{ kips}$$

For three spot arc welds

$$3\phi P_n = 19.23 \text{ kips} > 12.15 \text{ kips}.$$

Check Sheet-Tearing Capacity
Section E2.2.1

$$d_a = d - t = 1.0 - 0.1 = 0.9 \text{ in.}$$

$$\frac{d_a}{t} = \frac{0.9}{0.1} = 9$$

$$0.815\sqrt{\frac{E}{F_u}} = 0.815\sqrt{\frac{29500}{60}} = 18.1 > 9$$

Hence

$$P_n = 2.20td_aF_u \qquad\qquad \text{(Eq. E2.2.1-2)}$$
$$= 2.2 \times 0.1 \times 0.9 \times 60 = 11.88 \text{ kips}$$

Now $\phi = 0.60$ for Eq. (E2.2.1-2).

$$\phi P_n = 0.60 \times 11.88 = 7.13 \text{ kips}$$

For three spot arc welds

$$3\phi P_n = 21.4 \text{ kips} > 12.15 \text{ kips}$$

Minimum-Edge Distance and Spot Spacing
Section E2.2.1

$$e_{\min} = 1.5 \text{ in.} = 1.5d$$

Clear distance between welds $= 1$ in. $= 1.0d$.

$$e_{\min} = \frac{P_u}{\phi F_u t} \qquad\qquad \text{(Eq. 2.2.1-6b)}$$

Hence

$$P_u = e_{\min}\phi F_u t$$

Now $\phi = 0.70$ for Eq. (E2.2.1-6b).

Since $F_u/F_{sy} = 1.33 > 1.08$

$$P_u = 1.5 \times 0.70 \times 60 \times 0.1 = 6.3 \text{ kips}$$

For three arc spot welds

$$3P_u = 18.9 \text{ kips} > 12.15 \text{ kips}$$

Use 1.5-in. edge distance from the center of the weld to the edge of the plate and 2 in. between weld centers. The final detail is shown in Figure 9.17c, and, hence, 1.5-in. edge distance is satisfactory.

E. Arc Seam Weld Design (LRFD Method)
Use same width ($d = 1.0$ in.) as for the arc spot weld; hence $T_d = 12.15$ kips remains the same based on section design strength in tension.

Calculate Length Based on Sheet-Tearing Capacity
  Section E2.3

$$P_n = 2.5tF_u(0.25L + 0.96d_a) \qquad \text{(Eq. E2.3-2)}$$

Now $\phi = 0.60$ for Eq. (E2.3-2).
Hence

$$\phi P_n \geqslant T_d = 12.15 \text{ kips}$$
$$L = 4\left(\frac{T_d}{\phi(2.5tF_u)} - 0.96d_a\right)$$
$$= 4\left[\frac{12.15}{0.60 \times 2.5 \times 0.1 \times 60} - 0.96 \times 0.9\right]$$
$$= 1.94 \text{ in.}$$

Hence use $L = 2$ in.

Check Shear Capacity of Spot Weld

Section E2.3

$$P_n = \left[\pi\frac{d_e^2}{4} + Ld_e\right]0.75F_{xx} \qquad \text{(Eq. E2.3-1)}$$

$$= \left[\pi\left(\frac{0.55}{2}\right)^2 + (2 \times 0.55)\right]0.75 \times 60$$

$$= 60.2 \text{ kips}$$

Now $\phi = 0.60$ for Eq. (E2.3-1).

Hence

$$\phi P_n = 0.60 \times 60.2 = 36.1 \text{ kips} > T_d = 12.15 \text{ kips}$$

Check Minimum-Edge Distance

Section E2.3 (refers to Section E2.2-1)

$$e_{\min} = 3 \text{ in.} > 1.5d = 1.5 \text{ in.}$$

$$e_{\min} = \frac{P_u}{\phi F_{ut}} \qquad \text{(Eq. E2.2.1-6b)}$$

Hence

$$P_u = e_{\min}\phi F_{ut}$$

Now $\phi = 0.70$ since

$$\frac{F_u}{F_{sy}} = 1.33 > 1.08$$

$$P_u = 3 \times 0.70 \times 60 \times 0.1$$

$$= 12.6 \text{ kips} > 12.15 \text{ kips}$$

Hence, 3 in. edge distance is satisfactory, and the final detail is shown in Figure 9.17d.

9.8.2 Bolted Connection Design Example

Problem

Design a bolted connection in single shear to fully develop the strength of the net section of the sheet in Example 9.8.1, using bolts in the line of action of the force as shown in Figure 9.9b. Use $\frac{1}{2}$-in. bolts to ASTM A325 with washers under both head and nut. The example uses the ASD method. The following section numbers refer to the AISI Specification.

Solution

A. Plate Strength for Net Plate (ASD Method)
    Section C2

$$d = \tfrac{1}{2} \text{ in.}$$

$$d_h = \text{diameter of oversize hole}$$
$$= d + \tfrac{1}{8} = 0.625 \text{ in.}$$

Section C2

$$A_n = (b - d_h)t = (3 - 0.625)0.1 = 0.2375 \text{ in.}^2$$

T_n is lesser of

$$T_n = A_g F_y = (3 \times 0.1) \times 45 = 13.5 \text{ kips} \qquad \text{(Eq. C2-1)}$$

Since $\Omega = 1.67$ for Eq. (C2-1),

$$\frac{T_n}{\Omega} = 8.08 \text{ kips}$$

$$T_n = A_n F_u = 0.2375 \times 60 = 14.25 \text{ kips} \qquad \text{(Eq. C2-2)}$$

Since $\Omega = 2.00$ for Eq. (C2-2),

$$\frac{T_n}{\Omega} = 7.125 \text{ kips}$$

Hence, the required design strength is the lesser of 7.125 kips and 8.08 kips.

$$T_d = 7.125 \text{ kips}$$

Section E3.2
Where washers are provided under both the bolt head and the nut,

$$P_n = \left(1.0 - 0.9r + \frac{3rd}{s}\right) F_u A_n \leqslant F_u A_n \qquad \text{(Eq. E3.2-1)}$$

Now from Figure 9.9b, $s = 3$ in., $r = \frac{1}{3}$, assuming three bolts. Hence

$$P_n = \left[1.0 - \frac{0.9}{3} + 3\left(\frac{1}{3}\right)\frac{0.5}{3} \right] \times 60 \times 0.2375$$
$$= 12.35 \text{ kips} < F_u A_n = 14.25 \text{ kips}$$

Now $\Omega = 2.22$ for single shear connections in Eq. (E3.2-1).

$$\frac{P_n}{\Omega} = \frac{12.35}{2.22} = 5.56 \text{ kips} < 7.125 \text{ kips}$$

B.   Number of Bolts Required (ASD Method)

ASTM A325 bolt $F_{nv} = 54.0$ (when threads are not excluded from shear plane)

Section E3.4

$$A_b = \frac{\pi d^2}{4} = \frac{\pi(0.5)^2}{4} = 0.196 \text{ in.}^2$$
$$P_n = A_b F_{nv} = 0.196 \times 54.0 = 10.58 \text{ kips} \qquad \text{(Eq. E3.4-1)}$$

Now

$$\Omega = 2.4$$
$$\frac{P_n}{\Omega} = 4.41 \text{ kips} \qquad\qquad\qquad \text{(Table E3.4-1)}$$
$$\frac{3P_n}{\Omega} = 13.23 \text{ kips} > T_d = 5.56 \text{ kips}$$

C.   Check Bearing Capacity (ASD Method)

Section E3.3   Table E3.3-1 (with washers under both bolt head and nut)

For single shear

$$P_n = 3.00 F_u dt = 3.00 \times 60 \times 0.5 \times 0.1 = 9 \text{ kips}$$

Now

$$\Omega = 2.22$$

$$\frac{P_n}{\Omega} = 4.05 \text{ kips} \qquad \text{(Table E3.4-1)}$$

$$3\frac{P_n}{\Omega} = 12.16 \text{ kips} > T_d = 5.56 \text{ kips}$$

D. Edge Distance (ASD Method)
    Section E3.1
Use $e = 1.0$ in. $> 1.5d = 0.75$ in.

$$P_n = teF_u = 0.1 \times 1.0 \times 60 = 6 \text{ kips} \qquad \text{(Eq. E3.1-1)}$$

Now

$$\Omega = 2.0 \qquad \text{for } \frac{F_u}{F_{sy}} = 1.33 > 1.08$$

$$\frac{P_n}{\Omega} = 3.0 \text{ kips}$$

$$3\frac{P_n}{\Omega} = 9.0 \text{ kips} > 5.56 \text{ kips}$$

Also distance between center of bolt holes must be greater than or equal to $3d = 1.5$ in. $> e + \frac{1}{4}$ in. $= 1.25$ in. Hence, bolt hole spacing is governed by the $3d$ requirement and not tearout.

Final solution is three $\frac{1}{2}$-in. ASTM A325 bolts in line spaced 1.5 in. between the centers of the bolt holes and 1.0 in. from the end of the plate to the center of the last bolt hole. Allowable design load is 5.56 kips, which is controlled by the plate strength and not the bolt in shear or bearing strength.

9.8.3 Screw Fastener Design Example (LRFD Method)

Problem

Determine the design tension strength of a screwed connection where a No. 12 screw fastens trapezoidal sheeting of tensile strength (F_u) equal to 80 ksi and base metal thickness (BMT) of 0.0165 in. to a purlin of ASTM A570 Grade 50 steel and thickness 0.60 in.

Solution

Pullout Screw

Section E4.4.1

No. 12 screw:

$$d = 0.21 \text{ in.}$$

$F_{u2}(\text{purlin}) = 65 \text{ ksi}$

$$P_{not} = 0.85 t_c d F_{u2}$$
$$= 0.85 \times 0.060 \times 0.21 \times 65 \quad \text{(Eq. E4.4.4.1)}$$
$$= 0.696 \text{ kips}$$
$$\phi = 0.5 \quad \text{Section E4}$$
$$\phi P_{not} = 0.5 \times 0.696 = 0.348 \text{ kips}$$

Static Capacity of Sheathing Limited by Pullover

Section E4.4.2

$$P_{nov} = 1.5 t_1 d_w F_{u1} \qquad\qquad \text{(Eq. 4.4.2.1)}$$
$$t_1 = 0.0165 \text{ in.}$$
$$F_{u1} = 80 \text{ ksi}$$
$$d_w = 0.375 \text{ in.}$$
$$P_{nov} = 1.5 \times 0.0165 \times 0.375 \times 80 = 0.742 \text{ kips}$$
$$\phi = 0.5 \quad \text{for Section E4}$$
$$\phi P_{nov} = 0.5 \times 0.742 = 0.371 \text{ kips}$$

Hence, pullout rather than pullover controls the design.

REFERENCES

9.1 Peköz, T. and McGuire, W., Welding of Sheet Steel, Proceedings, Fifth International Specialty Conference on Cold-Formed Steel Structures, St Louis, MO, November, 1980.

9.2 Stark, J. W. B. and Soetens, F., Welded Connections in Cold-Formed Sections, Fifth International Specialty

Conference on Cold-Formed Steel Structures, St Louis, MO, November, 1980.

9.3 American Welding Society, ANSI/AWS D1.3-98, Structural Welding Code—Sheet Steel, American Welding Society, Miami, FL, 1998.

9.4 American Welding Society, AWS C1.3-70, Recommended Practises for Resistance Welding Coated Low-Carbon Steels, Miami, FL, 1970 (reaffirmed 1987).

9.5 American Welding Society, AWS C1.1-66, Recommended Practice for Resistance Welding, Miami, FL.

9.6 Yu, Wei-Wen, AISI Design Criteria for Bolted Connections, Proceedings Sixth International Specialty Conference on Cold-Formed Steel Structures, St Louis, MO, November, 1982.

9.7 Rogers, C. A and Hancock, G. J., Bolted connection tests of thin G550 and G300 sheet steels, Journal of Structural Engineering, ASCE, Vol. 124, No. 7, 1998, 798–808.

9.8 European Convention for Constructional Steelwork, European Recommendations for the Design of Light Gauge Steel Members, Technical Committee 7, Working Group 7.1, 1987.

9.9 Stark, J. W. B. and Toma, A. W., Connections in thin-walled structures, Developments in Thin-Walled Structures, Chapter 5, Eds. Rhodes, J. and Walker, A. C., Applied Science Publishers, London, 1982.

9.10 Peköz. T., Design of Cold-Formed Steel Screw Connections, Tenth International Specialty Conference on Cold-Formed Steel Structures, St Louis, November, 1990.

9.11 Birkemoe, P. C. and Gilmor, M. I., Behaviour of bearing-critical double angle beam connections, Engineering Journal, AISC, Fourth Quarter, 1978.

9.12 American Institute of Steel Construction, Specification for the Design, Fabrication and Erection of Structural Steel for Buildings, Chicago, AISC, 1978.

10

Metal Building Roof and Wall Systems

10.1 INTRODUCTION

Metal building roof and wall systems are generally constructed with cold-formed C- and Z-sections. Roof members are called *purlins*, and wall members are referred to as *girts*, although the sections may be identical. Generally, purlins are lapped, as shown in Figure 10.1, to provide continuity and, therefore, greater efficiency. The lap connection is usually made with two machine-grade bolts through the webs of the lapped purlins at each end of the lap as shown in the figure. In addition, the purlins are flange-bolted to the supporting rafter or web-bolted to a web clip or an antiroll device. Z-section purlins are generally point symmetric; however, some manufacturers produce Z-sections with unequal-width flanges to facilitate nesting in the lapped region. Girts are generally not lapped but can be face- or flush-mounted as shown in Figure 10.2.

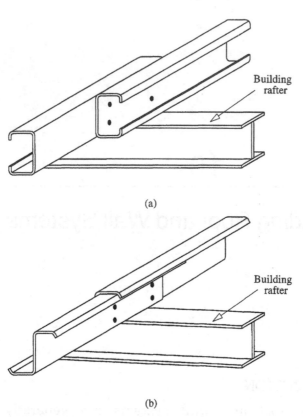

(a)

(b)

FIGURE 10.1 Continuous or lapped purlins: (a) lapped C-sections; (b) lapped Z-sections

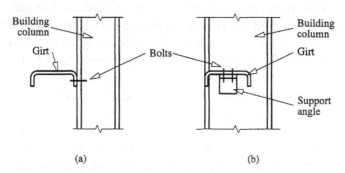

(a)   (b)

FIGURE 10.2 Face- and flush-mounted girts: (a) face mounted girt; (b) flush mounted girt.

Metal building roof systems are truly structural systems: the roof sheathing supports gravity and wind uplift loading and provides lateral support to the purlins. In turn, the purlins support the roof sheathing and provide lateral support, by way of flange braces, to the supporting main building frames. In addition, purlins may be required to carry axial force from the building end walls to the longitudinal wind bracing system. These purlins are referred to as strut purlins. Figure 10.3 shows typical framing for a metal building.

Roof panels are one of two basic types: through- or screw-fastened and standing seam. Panel profiles are commonly referred to as pan-type (Figure 10.4a) or rib-type (Figure 10.4b). Through-fastened panels are attached directly to the supporting purlins with self-drilling or self-tapping screws (Figure 10.5a). Thermal movement of attached panels can enlarge the screw holes, resulting in roof leaks. However, through-fastened panels can provide full lateral support to the connected purlin flange.

Standing seam roofing provides a virtually penetration-free surface resulting in a watertight roof membrane. Except at the building eave or ridge, standing seam panels are attached to the supporting purlins with concealed clips which are screw-fastened to the supporting purlin flange (Figure 10.5b). There are two basic clip types: fixed (Figure

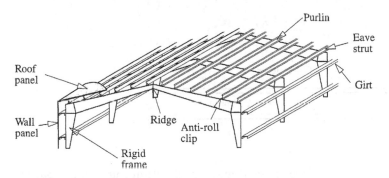

FIGURE 10.3 Typical metal building framing.

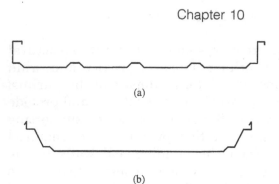

(a)

(b)

FIGURE 10.4 Roof panel profiles: (a) pan-type panel profile; (b) rib-type panel profile.

10.6a) and sliding or two-piece (Figure 10.6b). Thermal movement is accounted for by movement between the roof panel and the fixed clip or by movement between the parts of the sliding clips. The lateral support provided by standing seam panels and clips is highly dependent on the panel profile and clip details.

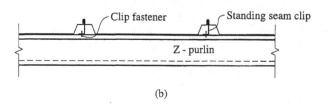

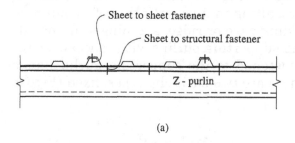

FIGURE 10.5 Through-fastened and standing seam panels: (a) through-fastened panel; (b) standing seam panel.

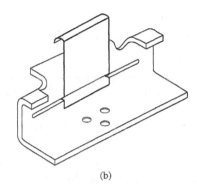

(a)   (b)

FIGURE 10.6 Types of standing seam clips: (a) fixed clip;
(b) sliding or two-piece clip.

Wall sheathing is cold-formed in a large variety of
profiles with both through-fastened and fixed-clip fastening
systems. Design considerations for wall sheathing are
basically the same as for roof sheathing.

The effective lateral support provided by the
panel/attachment system is a function of the system details
as well as the loading direction. Generally, through-
fastened sheathing is assumed to provide continuous
lateral and torsional restraints for gravity loading in the
positive moment region (the portion of the span where the
panel is attached to the purlin compression flange). Design
assumptions for the negative moment region (the portion of
the span where the panel is attached to the purlin tension
flange) vary from unrestrained to fully restrained. A
common assumption is that the purlin is unbraced between
the end of the lap and the adjacent inflection point (Ref.
10.1). However, recent testing has shown that this assumption may be unduly conservative (Ref. 10.2).

For uplift loading, through-fastened sheathing
provides lateral, but not full torsional, restraint. Attempts
have been made to develop test methods to determine the
torsional restraint provided by specific panel profile/screw
combinations as described in Section 5.3. However, the

301

variability of the methods and their complexity necessitated something simpler for routine use. Consequently, the empirical R-factor method was developed for determining the flexural strength of through-fastened roof purlins under uplift loading. The method is described in Section 10.2.2.

The lateral and torsional restraints provided by standing seam roof systems vary considerably, depending on the panel profile and the clip details. Consequently, a generic solution is not possible. The so-called base test method was therefore developed; it is described in Section 10.2.3.

10.2 SPECIFIC AISI DESIGN METHODS FOR PURLINS

10.2.1 General

The AISI Specification provides empirical and test-based methods for determining the uplift loading strength of through-fastened panel systems (R-factor method) and for the gravity and uplift loading strength of standing seam panel systems (base test method). Determination of the gravity loading strength of through-fastened panel systems is not specified. The industry practice is to assume full lateral and torsional support in the positive moment region and no lateral support between an inflection point and the end of the lap in the negative moment region. However, recent testing (Ref. 10.2) has shown that full lateral and torsional support can also be assumed in this region.

Rational methods, such as that presented in Section 5.3, are permitted by the AISI Specification for certain circumstances. For instance, in Section C3.1.3, Beam Having One Flange Through-Fastened to Deck or Sheathing, it is stated that "if variables fall outside any of the above the limits, the user must perform full scale tests... or apply a rational analysis procedure."

10.2.2 R-Factor Design Method for Through-Fastened Panel Systems and Uplift Loading

The design procedure for purlins subject to uplift loading in Section C3.1.3 of the AISI Specification Supplement No. 1 is based on the use of reduction factors (R-factors) to account for the torsional-flexural or nonlinear distortional behavior of purlins with through-fastened sheathing. The R-factors are based on tests performed on simple span and continuous span systems. Both C- and Z-sections were used in the testing programs; all tests were conducted without intermediate lateral bracing (Ref. 10.3).

The R-factor design method simply involves applying a reduction factor (R) to the bending section strength $(S_e F_y)$, as given by Eq. (10.1), to give the nominal member moment strength:

$$M_n = R S_e F_y \tag{10.1}$$

with $R = 0.6$ for continuous span C-sections
$\quad\ = 0.7$ for continuous span Z-sections
$\quad\ =$ values from AISI Specification Supplement No. 1
$\quad\quad$ Table C3.1.3-1 for simple span C- and Z-sections.

The rotational stiffness of the panel-to-purlin connection is due to purlin thickness, panel thickness, fastener type and location, and insulation. Therefore, the reduction factors only apply for the range of sections, lap lengths, panel configurations, and fasteners tested as set out in Section C3.1.3 of the AISI Specification Supplement No. 1. For continuous span purlins, compressed glass fiber blanket insulation of thickness between 0 and 6 in. does not measurably affect the purlin strength (Ref. 10.4). The effect is greater for simple span purlins (Ref. 10.5), requiring that the reduction factor (R) be further reduced to rR, where

$$r = 1.00 - 0.01 t_i \tag{10.2}$$

303

with t_i = thickness of uncompressed glass fiber blanket insulation in inches.

The resulting design strength moment ($\phi_b M_n$) or allowable stress design moment capacity (M_n/Ω_b) is compared with the maximum bending moment in the span determined from a simple elastic beam analysis. The resistance factor (ϕ_b) is 0.90, and the factor of safety (Ω_b) is 1.67.

The AISI Specification R-factor design method does not apply to the region of a continuous beam between an inflection point and a support nor to cantilever beams. For these cases, the design must consider buckling as described in Section 5.2. If variables are outside of the Specification limits, full-scale tests or a rational analysis may be used to determine the design strength.

10.2.3 The Base Test Method for Standing Seam Panel Systems

The lateral and torsional restraints provided by standing seam sheathing and clips depend on the panel profile and clip details. Lateral restraint is provided by friction in the clip and drape or hugging of the sheathing. Because of the wide range of panel profiles and clip details, a generic solution for the restraint provided by the system is impossible. For this reason, the base test method was developed by Murray and his colleagues at Virginia Tech (Refs. 10.6–10.10). The method uses separate sets of simple span, two-purlin-line tests to establish the nominal moment strength of the positive moment regions of gravity-loaded systems and the negative moment regions of uplifted loaded systems. The results are then used to predict the strength of multispan, multiline systems for either gravity or wind uplift loadings.

The base test method was verified by 11 sets of gravity loading tests and 16 sets of wind uplift loading tests. Each set consisted of one two-purlin-line simple span test and one two- to four-purlin line, two or three continuous span test. Failure loads for the multiple-span tests were predicted using results from the simple span tests for the positive moment region strength and AISI Specification provisions for the negative moment region strength. The procedure for determining the predicted failure load of the continuous span test using the results of the single-span base test is illustrated in Figure 10.7 for gravity loading.

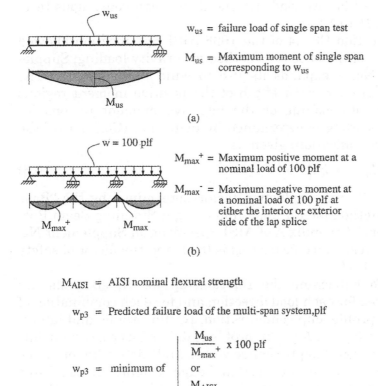

w_{us} = failure load of single span test

M_{us} = Maximum moment of single span corresponding to w_{us}

(a)

M_{max}^+ = Maximum positive moment at a nominal load of 100 plf

M_{max}^- = Maximum negative moment at a nominal load of 100 plf at either the interior or exterior side of the lap splice

(b)

M_{AISI} = AISI nominal flexural strength

w_{p3} = Predicted failure load of the multi-span system, plf

$$w_{p3} = \text{minimum of} \begin{vmatrix} \dfrac{M_{us}}{M_{max}^+} \times 100 \text{ plf} \\ \text{or} \\ \dfrac{M_{AISI}}{M_{max}^-} \times 100 \text{ plf} \end{vmatrix}$$

(c)

FIGURE 10.7   Verification of base test method: (a) single-span test case; (b) multi-span stiffness analysis; (c) predicted failure load.

The predicted failure load is then compared to the experimental failure load of the continuous span test.

The gravity loading series included tests with lateral restraint only at the rafter supports and tests with lateral restraint at the rafter supports plus intermediate third-point restraint. The ratio of actual-to-predicted failure loads for the three-span continuous tests was 0.87 to 1.02. The uplift loading tests were also conducted with rafter and third-point lateral restraints. The ratio of actual-to-predicted failure loads for the three-span continuous tests was 0.81 to 1.25.

Section C3.1.4 of the 1996 AISI Specification allowed the use of the base test method for gravity loading; Supplement No. 1 expands its use to wind uplift loading. The nominal moment strength of the positive moment regions for gravity loading or the negative moment regions for uplift loading is determined by using Eq. (C3.1.4-1) of the AISI Specification, given as

$$M_n = RS_eF_y \qquad (10.3)$$

where R is the reduction factor determined by the "Base Test Method for Purlins Supporting a Standing Seam Roof System," Appendix A of AISI Specification Supplement No. 1. The resistance factor (ϕ_b) is 0.90, and the factor of safety (Ω_b) is 1.67.

To determine the relationship for R, six tests are required for each load direction and for each combination of panel profile, clip configuration, purlin profile, and lateral bracing layout. A purlin profile is defined as a set of purlins with the same depth, flange width, and edge stiffener angle, but with varying thickness and edge stiffener length. Three of the tests are conducted with the thinnest material and three with the thickest material used by the manufacturer for the purlin profile. Components used in the base tests must be identical to those used in the actual systems. Generally, the purlins are oriented in the same direction (purlin top flanges facing toward the building ridge or toward the building eave) as used in the actual building. However, it is permitted to test with the purlin top flanges facing in opposite directions if additional analysis is done, even though that will not be the orientation in the actual building. The required analysis uses diaphragm test results.

Results from the six tests are then used in Eq. (10.4) to determine an R-factor relationship:

$$R = \left(\frac{R_{t\,\text{max}} - R_{t\,\text{min}}}{\overline{M}_{nt\,\text{max}} - \overline{M}_{nt\,\text{min}}} \right)(M_n - \overline{M}_{nt\,\text{min}}) + R_{t\,\text{min}} \leq 1.0$$

(10.4)

with $R_{t\,\text{min}}$, $R_{t\,\text{max}}$ = mean minus one standard deviation of the modification factors for the thinnest and thickest purlins tested, respectively. The values may not be taken greater than 1.0.

M_n = nominal flexural strength ($S_e F_y$) for the section for which R is being determined.
$\overline{M}_{nt_{\text{min}}}$, $\overline{M}_{nt_{\text{max}}}$ = average tested flexural strength for the thinnest and thickest purlins tested, respectively.

The reduction factor for each test (R_t) is computed from

$$R_t = \frac{M_{ts}}{M_{nt}}$$

(10.5)

with M_{ts} = maximum single-span failure moment from the test
M_{nt} = flexural strength calculated using the measured cross section and measured yield stress of the purlin.

Reduction factor values generally are between 0.40 and 0.98 for gravity and uplift loading, depending on the panel profile and clip details. For some standing seam Z-purlin systems, the uplift loading reduction factor is greater than the corresponding gravity loading reduction factor. Gravity loading tends to increase purlin rotation as shown in Figure 10.8a, and uplift loading tends to decrease Z-purlin rotation, as shown in Figure 10.8b. If sufficient torsional restraint is provided by the panel/clip connection, a greater strength may be obtained for uplift loading than for gravity loading.

The maximum single-span failure moment (M_{nt}) is determined by using the loading at failure determined from

$$w_{ts} = (p_{ts} + p_d)s + 2P_L\left(\frac{d}{B}\right) \tag{10.6}$$

with

$$P_L = 0.041\frac{b^{1.5}}{d^{0.90}t^{0.60}}(p_{ts} + p_d)s$$

and where p_{ts} = maximum applied load (force/area) in the test

p_d = weight of the purlin and roof panel (force/area), positive for gravity loading and negative for uplift loading

s = tributary width for the tested purlins

b = purlin flange width

d = depth of purlin

t = purlin thickness

B = purlin spacing.

The force P_L accounts for the effect of the overturning moment on the system due to anchorage forces. It applies only to Z-purlin gravity loading tests without intermediate lateral restraint, when the top flanges of the purlins point in the same direction and when the eave (or "downhill") purlin fails before the ridge (or "uphill") purlin.

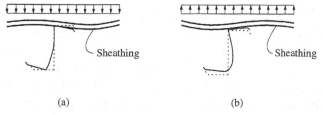

(a) (b)

FIGURE 10.8 Purlin rotation due to gravity and uplift loading: (a) gravity loading; (b) uplift loading.

The AISI base test procedure requires that the tests be conducted with a test chamber capable of supporting a positive or negative internal pressure differential. Figure 10.9 shows a typical chamber. Construction of a test setup must match that of the field erection manuals of the manufacturer of the standing seam roof system. The lateral bracing provided in the test must match the actual field conditions. For example, if antiroll devices are installed at the rafter support of each purlin in the test, then antiroll devices must be provided at every purlin support location in the actual roof. Likewise, if intermediate lateral support is provided in the test, the support configuration is required in the actual roof. If antiroll clips are used at the supports of only one purlin in the test, the unrestrained purlin can be considered as a "field" purlin as long as there is no positive connection between the two purlins. That is, if standing seam panels or an end-support angle are not screw-connected to both purlins, the actual "field" purlins

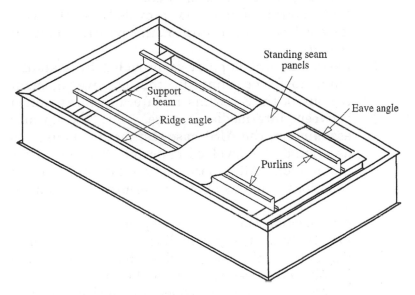

FIGURE 10.9   Base test chamber.

need only be flange-attached to the primary support member.

Example calculations for the determination of the R-relationship and its application are in Section 10.5.1.

10.3 CONTINUOUS PURLIN LINE DESIGN

Because Z-purlins are point symmetric and the applied loading is generally not parallel to a principal axis, the response to gravity and wind uplift loading is complex. The problem is somewhat less complex for continuous C-purlins since bending is about principal axes. Design is further complicated when a standing seam roof deck, which may provide only partial lateral restraint, is used. In addition to AISI Specification provisions, metal building designers use design and analysis assumptions, such as the following:

1. Constrained bending—that is, bending is about an axis perpendicular to the web.
2. Full lateral support is provided by through-fastened roof sheathing in the positive moment regions.
3. Partial lateral support is provided by standing seam roof sheathing in the positive moment regions, or the Z-purlins are laterally unrestrained between intermediate braces. For the former assumption, the AISI base test method is used to determine the level of restraint. For the latter assumption, AISI Specification lateral buckling equations are used to determine the purlin strength.
4. An inflection point is a brace point.
5. For analysis, the purlin line is considered prismatic (e.g., the increased stiffness due to the doubled cross section within the lap is ignored), or the purlin line is considered nonprismatic (e.g., considering the increased stiffness within the lap).

6. Use of vertical slotted holes, which facilitate erection of the purlin lines, for the bolted lap web-to-web connection does not affect the strength of continuous purlin lines.

7. The critical location for checking combined bending and shear is immediately adjacent to the end of the lap in the single purlin.

Constrained bending implies that bending is about an axis perpendicular to the Z-purlin web and that there is no purlin movement perpendicular to the web. That is, all movement is constrained in a plane parallel to the web. Because a Z-purlin is point symmetric and because the applied load vector is not generally parallel to a principal axis of the purlin, the purlin tends to move out of the plane of the web, as discussed in Section 10.4.1. Constrained bending therefore is not the actual behavior. However, it is a universally used assumption and its appropriateness is even implied in the AISI Specification. For instance, the nominal strength equations for Z-sections in Section C3.1.2, Lateral Buckling Strength, of the 1996 AISI Specification apply to "Z-sections bent about the centroidal axis perpendicular to the web (x-axis)." All of the analyses referred to in this section are based on the constrained bending assumption.

It is also assumed that through-fastened roof sheathing provides full lateral support to the purlin in the positive moment region. However, it is obvious that this assumption does not apply equally to standing seam roof systems. The degree of restraint provided depends on the panel profile, seaming method, and clip details. The restraint provided by the standing seam system consists of panel drape (or hugging) and clip friction or lockup. Lower values are obtained when "snap together" (e.g., no field seaming) panels and two-piece (or sliding) clips are used. Higher values are obtained when field-seamed panels and fixed clips are used. However, exceptions apply to both of these general statements.

The 1996 AISI Specification in Section C3.1.4, Beams Having One Flange Fastened to a Standing Seam Roof System, allows the designer to determine the design strength of Z-purlins using (1) the theoretical lateral-torsional buckling equations in Section C3.1.2 or (2) the base test method described in Section 10.2.3. The base test method indirectly establishes the restraint provided by a standing seam panel/clip/bracing combination.

If intermediate braces are not used, a lateral-torsional buckling analysis predicts an equivalent R-value in the range 0.12–0.20. The corresponding base-test-generated R-value will be three to five times larger, which clearly shows the beneficial effects of panel drape and clip restraint. However, if intermediate bracing is used, base-test-method-generated R-values will sometimes be less than that predicted by a lateral-torsional analysis with the unbraced length equal to the distance between intermediate brace locations. The latter results are somewhat disturbing in that panel/clip restraint is not considered in the analytical solution, yet the resulting strength is greater than the experimentally determined value. Possible explanations are that the intermediate brace anchorage in the base test is not sufficiently rigid, that the Specification equations are unconservative, or that distortional buckling contributes to the failure mechanism (Ref. 10.11). The AISI Commentary to Section C3.1.2 states for Z-sections that "a conservative design approach has been and is being used in the *Specification*, in which the elastic critical moment is taken to be one-half of that of for I-beams," but no reference to test data is given.

One of two analysis assumptions are commonly made by designers of multiple-span, lapped, purlin lines: (1) prismatic (uniform) moment of inertia or (2) nonprismatic (nonuniform) moment of inertia. For the prismatic assumption, the additional stiffness caused by the increased moment of inertia within the lap is ignored. For the nonprismatic assumption, the additional stiffness is accounted for by using the sum of the moments of inertia of the purlins forming the lap. Larger positive moments and smaller negative moments result when the first assumption is used with gravity loading. The reverse is true for the second assumption. For uplift loading the same conclusions apply except that positive and negative moment locations are reversed. It follows then that the prismatic assumption is more conservative if the controlling strength location is in the positive moment region and that the nonprismatic assumption is more conservative if the controlling strength location is in the negative moment region, i.e., within or near the lap.

Since the purlins are not continuously connected within the lap, full continuity will not be achieved. However, the degree of fixity is difficult to determine experimentally. A study by Murray and Elhouar (Ref. 10.12) seems to indicate that the nonprismatic assumption is a more accurate approach. A more recent study by Bryant and Murray (Ref. 10.2) confirmed this result.

Murray and Elhouar (Ref. 10.12) analyzed the results of through-fastened panel multiple-span, Z-purlin-line, gravity-loaded tests conducted by 10 different organizations in the United States. The nonprismatic assumption was used to determine the analytical moments for the comparisons. Of the 24 (3 two-span and 21 three-span) continuous tests analyzed, the predicted critical limit state for 18 tests was combined bending and shear at the end of the lap in the exterior span and for two tests it was positive moment failure in the exterior span. The predicted critical location for the remaining three tests was at the end of the lap in the interior span. The average experimental-to-predicted ratio when combined bending and shear controlled was 0.93 with a range of 0.81–1.06. The corresponding ratios for the three tests when positive moment controlled were 0.93 and 0.94. If the prismatic assumption had been used to determine moments from the experimental failure load, the combined bending and shear ratios would have been somewhat less and the positive moment ratios somewhat higher. A ratio less than one indicates that the predicted value is less than the experimental value or an unconservative result; therefore the prismatic assumption would be more unconservative.

Seven two- and three-span continuous, simulated gravity loading, two-Z-purlin-line tests were conducted by Bryant and Murray (Ref. 10.2) in which strain gauges were installed on the tension flange at and near the theoretical inflection point of one exterior span. The location of the inflection point was determined under the nonprismatic assumption. Three of the tests were conducted with through-fastened sheathing and four with standing seam sheathing. Figure 10.10 shows typical applied load versus measured tension flange strain results. Position 8 is at the theoretical inflection point location (I.P.), positions 6 and 10 are 12 in. each side of the I.P. location, and positions 7 and 9 are 6 in. each side of the I.P. The initial strains at position 8 are essentially zero and were not above 150 microstrain by

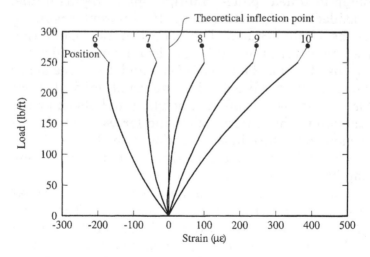

FIGURE 10.10   Load vs. measured strain near a theoretical inflection point.

the end of the test. From these studies, it is apparent that non-prismatic analysis is appropriate for the design of continuous Z-purlin lines.

For many years it has been generally accepted that an inflection point is a brace in a Z-purlin line. Work by Yura (Ref. 10.13) on H-shapes led him to conclude that an inflection point is not a brace point unless both lateral and torsional movement are prevented. The 1996 AISI Specification is silent on this issue; however, the expression for the bending coefficient, C_b (Eq. 5.16), which is dependent on the moment gradient as shown in Figure 5.7, is included. This expression was adopted from the AISC Load and Resistance Factor Design (LRFD) Specification (Ref. 10.14). The Commentary to this specification states that an inflection point cannot be considered a brace point. In the AISI Design Guide, A Guide For Designing with Standing Seam Roof Panels (Ref. 10.1), the inflection point is not considered as a brace point and C_b is taken as 1.0. In Ref. 10.15, the inflection point is considered a brace point and C_b is taken as 1.75.

Because C- and Z-purlins tend to rotate or move in opposite directions on each side of an inflection point, tests were conducted by Bryant and Murray (Ref. 10.2) to determine if in fact an inflection point is a brace point in continuous, gravity-loaded, C- and Z-purlin lines of through-fastened and standing seam systems. In the study, instrumentation was used to verify the actual location of the inflection point as described above and the lateral movement of the bottom flange of the purlins on each side of the inflection point, as well as near the maximum location in an exterior span. The results were compared to movement predicted by finite element models of two of the tests. Both the experimental and analytical results showed that, although lateral movement did occur at the inflection point, the movement was considerably less than at other locations along the purlins. The bottom flanges on both sides of the inflection point moved in the same direction, and double curvature was not apparent from either the experimental or finite element results. The lateral movement in the tests using lapped C-purlins was larger than the movement in the Z-purlin tests, but was still relatively small. From these results it would seem that an inflection point is not a brace point.

The predicted and experimental controlling limit state for the three tests using through-fastened roof sheathing was shear plus bending failure immediately outside the lap in the exterior test bay. The experiment failure loads were compared to predicted values using provisions of the AISI Specifications and assuming (1) the inflection point is not a brace point, (2) the inflection point is a brace point, and (3) the negative moment region strength is equal to the effective yield moment, $S_e F_y$. All three analysis assumptions resulted in predicted failure loads less than the experimental failure loads: up to 23% for assumption (1), up to 11% for assumption (2), and approximately 8% for assumption (3). It is difficult to draw definite conclusions from this data. However, it is clear that the bottom flange of a continuous purlin line moves laterally in the same direction on both sides of an inflection point but the movement is relatively small. It appears that even the assumption of full lateral restraint for through-fastened roof systems is conservative.

The web-to-web connection in lapped Z-purlin lines is generally two $\frac{1}{2}$-in.-diameter machine bolts approximately $1\frac{1}{2}$ in. from the end of the unloaded purlin as shown in Figure 10.1. To facilitate erection, vertical slotted web holes are generally used, which may allow slip in the lap invalidating the continuous purlin assumption. Most, if not all, of the continuous span tests referred to above used purlins with vertical slotted holes. There is no indication in the data that the use of slots in the web connections of lapped purlins has any effect on the strength of the purlins.

Both the shear and moment gradients between the inflection point and rafter support of continuous purlin lines are steep. As a result, the location where combined bending and shear is checked can be critical. The metal building industry practice is to assume the critical location is immediately outside of the lapped portions of continuous Z-purlin systems—that is, in the single purlin, as opposed to at the web bolt line. The rationale for the assumption is that, for cold-formed Z-purlins, the limit state of combined bending and shear is actually web buckling. Near the end of the lap and, especially, at the web-to-web bolt line, out-of-plane movement is restricted by the nonstressed purlin section; thus buckling cannot occur at this location. Figure 10.11 verifies this contention. The corresponding assumption for C-purlin systems is that the shear plus bending limit state occurs at the web-to-web vertical bolt line.

The design of a continuous Z-purlin line is illustrated in Section 10.5.2.

FIGURE 10.11   Photograph of failed purlin at end of lap.

10.4 SYSTEM ANCHORAGE REQUIREMENTS

10.4.1 Z-Purlin Supported Systems

The principal axes of Z-purlins are inclined from the center-line of the web at an angle θ_p, as shown in Figure 10.12a. For Z-purlins commonly used in North America, θ_p is between 20° and 30°. For an unrestrained purlin, the component of the roof loading that is parallel to the web causes a Z-purlin to deflect in the vertical and horizontal directions, as shown in Figure 10.12b. The component of the roof loading perpendicular to the purlin web decreases the deflection perpendicular to the web. For steep roofs this component can even cause movement in the opposite direction, as shown in Figure 10.12c. Since the bottom flange of the purlin is restrained at the rafter connection and the top flange is at least partially restrained by the roof deck, a purlin tends to twist or "roll" as shown in Figure 10.12d. Devices such as antiroll clips at the rafters or intermediate lateral braces are used to minimize purlin roll. The force

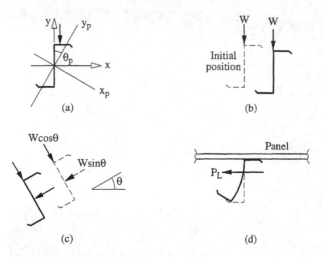

(a)

(b)

(c)

(d)

FIGURE 10.12 Z-purlin movement: (a) axes; (b) unrestrained movement; (c) movement because of large downslope component; (d) panel and anchorage restraints.

318

induced in these devices is significant and must be considered in design.

From basic principles (Ref. 10.16), the required anchorage or restraint force for a single Z-purlin with uniform load (W) parallel to the web is

$$P_L = 0.5\left(\frac{I_{xy}}{I_x}\right)W \tag{10.7}$$

where I_{xy} is the product moment of inertia and I_x is the moment of inertia with respect to the centroidal axis perpendicular to the web of the Z-section as shown in Figure 10.12a. For this formulation, the required anchorage force for a multi-purlin-line system is directly proportional to the number of purlins (n_p). However, Elhouar and Murray (Ref. 10.17) showed that the anchorage force predicted by Eq. (10.7) is conservative because of a system effect. Figure 10.13 shows a typical result from their study. The straight line is from Eq. (10.7), and the curved line is from stiffness analyses of a multi-purlin-line system. The difference in the anchorage force between the two results is a consequence of the inherent restraint in the system due to purlin web flexural stiffness and a Vierendeel truss action

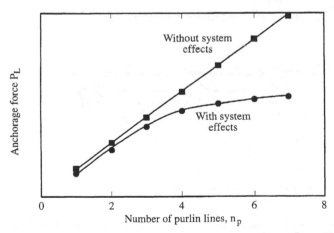

FIGURE 10.13 Anchorage forces vs. number of purlin lines.

caused by the interaction of purlin webs with the roof panel and rafter flanges. This Vierendeel truss action explains the relative decrease in anchorage force as the number of purlin lines (n_p) increases, as shown in the figure.

Section D3.2.1 of the AISI Specification has provisions that predict required anchorage forces in Z-purlin-supported roof systems supporting gravity loading and with all compression flanges facing in the same direction. The provisions were developed by using elastic stiffness models of flat roofs (Ref. 10.17) and were verified by full-scale and model testing (Ref. 10.18). For example, the predicted anchorage force in each brace (P_L) for single-span systems with restraints only at the supports, Figure 10.14, is

$$P_L = 0.5\beta W, \qquad \text{with } \beta = \frac{0.220b^{1.5}}{n_p^{0.72}d^{0.90}t^{0.60}} \qquad (10.8)$$

where W = applied vertical load (parallel to the web) for the set of restrained purlins
$\quad b$ = flange width
$\quad d$ = depth of section
$\quad t$ = thickness
$\quad n_p$ = number of restrained purlin lines

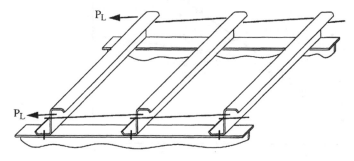

FIGURE 10.14 Single-span purlin system with anchorage at rafters.

The restraint force ratio, β, represents the system effect and was developed from regression analysis of stiffness model results.

Assuming that the top flange of the Z-purlin is in the upslope direction, the anchorage force P_L for single-span systems with restraints only at the rafter supports becomes

$$P_L = 0.5(\beta \cos \theta - \sin \theta)W \tag{10.9}$$

with $\theta =$ roof slope measured from the horizontal as shown in Figure 10.12c. The terms $W\beta \cos \theta$ and $W \sin \theta$ are the gravity load components parallel and perpendicular to the purlin web, respectively. The latter component is also referred to as the downslope component. Equations of this form are included in AISI Specification Supplement No. 1.

Two important effects were not taken into account in the development of Eq. (10.9). First, the internal system effect β applies to the forces $W \cos \theta$ and $W \sin \theta$. Second, the internal system effect reverses when the net anchorage force changes from tension to compression with increasing slope angle. Further, the stiffness models used to develop the AISI provisions assume a roof panel shear stiffness of 2500 lb/in., whereas the actual shear stiffness can be between 1000 lb/in. (standing seam sheathing) to as high as 60,000–100,000 lb/in. (through-fastened sheathing).

Neubert and Murray (Ref. 10.19) have recently proposed the following model, which eliminates the deficiencies inherent in Eq. (10.9). They postulate that the predicted anchorage force in any given system is equal to the anchorage force required for a single purlin (P_0) multiplied by the total number of purlins (n_p), a brace location factor (C_1), a reduction factor due to system effects (α), and modified by a factor for roof panel stiffness (γ). Thus,

$$P_L = P_0 C_1 (n_p^* \alpha + n_p \gamma) \tag{10.10}$$

Figure 10.15a shows a Z-purlin with zero slope and applied load W_p, which is the total gravity load acting on each purlin span:

$$W_p = wL \tag{10.11}$$

where $w =$ distributed gravity load on each purlin (force/length)
$L =$ span length

The fictitious force $W_p(I_{xy}/I_x)$ is the fictitious lateral force from basic principles (Ref. 10.16). The couple $W_p \delta b$ results from the assumed location of the load W_p on the flange—that is, δb from the face of the web as shown in Figure 10.15a. Figure 10.15b shows the set of real and fictitious forces associated with a single purlin on a roof with slope θ. The purlin is assumed to be pinned at the rafter connection. The set of forces accounts for the following effects: $W_p \sin \theta$ is the downslope component, $W_p \cos \theta(I_{xy}/I_x)$ is the downslope component of the fictitious lateral force, and $W_p \cos \theta(\delta b)$ is the net torque induced by eccentric loading of the top flange. Summation of moments about the pinned support results in

$$P_0 = \left[\left(\frac{I_{xy}}{2I_x} + \frac{\delta b}{d} \right) \cos \theta - \sin \theta \right] W_p \tag{10.12}$$

For low slopes, P_0 is positive (tension) and for high slopes P_0 is negative (compression). A positive value indicates that the purlin is trying to twist in a clockwise

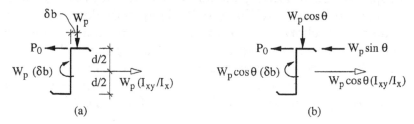

(a) (b)

FIGURE 10.15 Modeling of gravity loads: (a) forces for a single purlin on a flat roof; (b) forces for a single purlin on a sloping roof.

direction; a negative value means twist is in the counter-clockwise direction. The anchorage force is zero for

$$\theta_0 = \tan^{-1}\left(\frac{I_{xy}}{2I_x} + \frac{\delta b}{d}\right) \tag{10.13}$$

For $\theta < \theta_0$, P_0 is in tension, and for $\theta > \theta_0$, P_0 is in compression. The distance δb has a significant effect on the value of the intercept angle θ_0. In Ref. 10.19, δb was taken as $b/3$, based on anchorage forces measured in tests of zero-slope roof systems (Ref. 10.17).

Statistical analysis of stiffness model results was used to develop a relationship for the system effect factor α:

$$\alpha = 1 - C_2\left(\frac{t}{d}\right)(n_p^* - 1) \tag{10.14}$$

with C_2 a constant factor that depends on the bracing configuration and n_p^* is described below. The factor α is dimensionless and, when included in Eq. (10.10), models the reversal of the system effect when P_0 changes from tension to compression. The coefficient C_2 was determined from a set of regression analyses with values tabulated in Table 10.1. The value of the coefficient differs for each bracing configuration because bending resistance changes depending on the distance between a brace and rafter supports or other braces.

Equation (10.10) is quadratic with respect to n_p, because α is linear with n_p. For some value of n_p, denoted as $n_{p(\max)}$, P_L will reach a maximum and then decrease as n_p is increased above $n_{p(\max)}$. From basic calculus, $n_{p(\max)}$ is determined as

$$n_{p(\max)} = 0.5 + \frac{d}{2C_2 t} \tag{10.15}$$

However, the required anchorage force can never decrease as the number of purlins is increased. Therefore n_p^* is used in Eqs. (10.10) and (10.14) instead of n_p, where n_p^* is the

TABLE 10.1   Anchorage Force Coefficients for Use in Eq. (10.10)

Bracing configuration	C_1	C_2	C_3
Support anchorages only			
Single span	0.50	5.9	0.35
Multispan, exterior supports	0.50	5.9	0.35
Multispan, interior supports	1.00	9.2	0.45
Third-point anchorages			
Single span	0.50	4.2	0.25
Multispan, exterior spans	0.50	4.2	0.25
Multispan, interior spans	0.45	4.2	0.35
Midspan anchorages			
Single span	0.85	5.6	0.35
Multispan, exterior spans	0.80	5.6	0.35
Multispan, interior spans	0.75	5.6	0.45
Quarter-point anchorages			
Single span, outside $\frac{1}{4}$ points	0.25	5.0	0.35
Single span, midspan location	0.45	3.6	0.15
Multispan, exterior span $\frac{1}{4}$ points	0.25	5.0	0.40
Multispan, interior span $\frac{1}{4}$ points	0.22	5.0	0.40
Multispan, midspan locations	0.45	3.6	0.25
Third-point plus support anchorages			
Single span, at supports	0.17	3.5	0.35
Single span, interior locations	0.35	3.0	0.05
Multispan, exterior support	0.17	3.5	0.35
Multispan, interior support	0.30	5.0	0.45
Multispan, third-point locations	0.35	3.0	0.10

minimum of $n_{p(\max)}$ and n_p. In other words, P_L remains constant when the number of purlin lines exceeds $n_{p(\max)}$.

The brace location factor C_1 in Eq. (10.10) represents the percentage of total restraint that is allocated to each brace in the system. The sum of the C_1 coefficients for the braces in a span length is approximately equal to unity. The values for C_1 were determined from a regression analysis and are tabulated for each bracing configuration in Table 10.1. For multiple span systems, the C_1 values are larger for exterior restraints than the corresponding interior restraints, as expected from structural mechanics.

Figure 10.16 shows a typical plot comparing the Eq. (10.10) with, $\gamma = 0$, to the AISI Specification Eq. (10.9) with respect to slope angle θ. Equation (10.10) predicts slightly

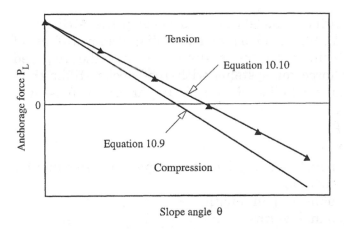

FIGURE 10.16 Anchorage force vs. roof slope: Eqs. (10.9) and (10.10).

greater anchorage force for low-slope roofs and significantly less anchorage force for high-slope roofs. The latter is due to the inclusion of system effects in the downslope direction which are ignored in the AISI provisions. Figure 10.17 shows a similar plot with respect to the number of restrained purlin lines.

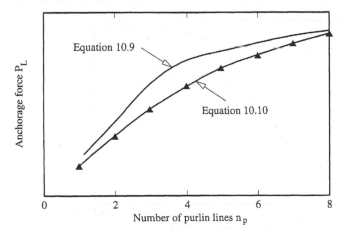

FIGURE 10.17 Anchorage force vs. number of purlins: Eqs. (10.9) and (10.10).

The stiffness models used to develop the AISI provisions, Eq. (10.9), included a roof deck diaphragm stiffness of $G' = 2500\,\text{lb/in.}$ with the assumption that the required anchorage force for systems with roof decks stiffer than $2500\,\text{lb/in.}$ is negligible. Roof deck diaphragm stiffness was taken as

$$G' = \frac{PL}{4a\Delta} \tag{10.16}$$

where $P =$ point load applied at midspan of a rectangular roof panel,
$\quad L =$ panel's span length
$\quad a =$ panel width

and

$\quad \Delta =$ deflection of the panel at the location of the point load.

In Ref. 10.19 stiffness models of roof systems with up to eight restrained purlins were analyzed, and it was found that deck stiffness above $2500\,\text{lb/in.}$ caused significant increases in the anchorage forces for systems with four or more purlin lines. The panel stiffness modifier (γ) in Eq. (10.10) accounts for this effect. As shown in Figure 10.18,

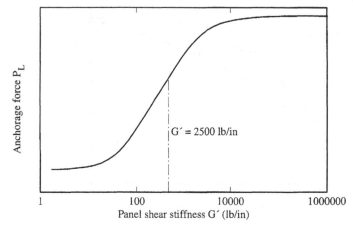

FIGURE 10.18 Anchorage force vs. panel stiffness (stiffness model).

the anchorage force varies linearly with the common logarithm of the roof panel shear stiffness over a finite range of stiffnesses, which leads to the following equation for the panel stiffness modifier:

$$\gamma = C_3 \log\left(\frac{G'}{2500\,\text{lb/in.}}\right) \qquad (10.17)$$

where G' = roof panel shear stiffness (lb/in.)

C_3 = constant from regression analysis of stiffness model

When $G' = 2500\,\text{lb/in.}$, $\gamma = 0$ as necessary, and γ is positive for $G' > 2500\,\text{lb/in.}$ and negative for $G' < 2500\,\text{lb/in.}$ Thus, when $G' > 2500\,\text{lb/in.}$, the anchorage force is increased, and when $G' < 2500\,\text{lb/in.}$, the required anchorage force is decreased. The values of C_3, which depend on the location of a brace relative to rafter supports and other braces, are tabulated for each bracing configuration in Table 10.1.

The effect of γ is to adjust the system effect factor, α. In Eq. (10.10) γ is multiplied by n_p instead of n_p^*, because as panel stiffness changes, change in anchorage force depends on the total number of purlins in the system and $n_{p(\text{max})}$ no longer applies. The panel stiffness modifier γ is valid only for $1000\,\text{lb/in.} \le G' \le 100{,}000\,\text{lb/in.}$ This is the range of linear behavior and most roof decks have shear stiffness within this limitation.

For very large panel stiffness, Eq. (10.10) can predict an anchorage force greater than that obtained from basic principles [Eq. (10.7)]. However, the maximum anchorage force is the anchorage force ignoring system effects. Thus,

$$|P_L| \le |P_0 C_1 n_p| \qquad (10.18)$$

Figure 10.19 is a typical plot of anchorage force versus panel stiffness for Eq. (10.10), compared with stiffness model results.

Since the stiffness models used to develop Eq. (10.10) had eight purlin lines or fewer, Eq. (10.10) must be used

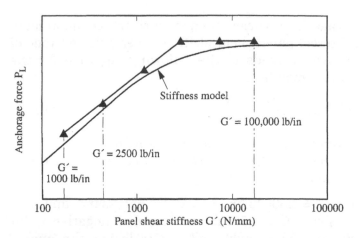

FIGURE 10.19 Anchorage force vs. Panel stiffness: Eq. (10.10) and stiffness model results.

with caution when there are more than eight purlin lines. If the top flanges of adjacent purlins or sets of purlins face in opposite directions, an anchorage is required to resist only the net downslope component.

Anchorage force calculations for a Z-purlin supported roof system are illustrated in Section 10.5.3.

10.4.2 C-Purlin Supported Systems

The anchorage requirements for C-purlins are much less than those for Z-purlins since bending is about a principal axis. The eccentricity caused by the load not acting through the shear center requires a resisting force of about 5% of the load parallel to the web of the C-section. AISI Specification Supplement No. 1 Section D3.2.1(a), C-Sections, gives the anchorage force for a C-section-supported roof system with all compression flanges facing in the same direction (Eq. D3.2.1-1) as

$$P_L = (0.05\alpha \cos \theta - \sin \theta)W \tag{10.19}$$

with $\alpha = +1$ for purlins facing in the upslope direction and $= -1$ for purlins facing in the downslope direction. A positive value of P_L means that anchorage is required to prevent movement of the purlin flanges in the upslope direction, and a negative value means that anchorage is required to prevent movement in the downslope direction.

As with Z-purlin systems, if the top flanges of adjacent C-purlin lines face in opposite directions, the anchorage need only resist the downslope component of the gravity load.

10.5 EXAMPLES

10.5.1 Computation of R-value from Base Tests

Problem

Determine the R-value relationship for the gravity loading base test data shown. The tests were conducted using Z-sections with nominal thicknesses of 0.060 in. and 0.095 in. and nominal yield stress of 55 ksi. The span length was 22 ft 9 in., and intermediate lateral braces were installed at the third points. The total supported load (w_{ts}) is equal to the sum of the applied load (w) and the weight of the sheathing and purlins (w_d). The maximum applied moment is M_{ts}.

Solution

A. Summary of Test Loadings

Test no.	Max. applied load w (lb/ft)	Deck weight (lb/ft)	Purlin weight (lb/ft)	w_d (lb/ft)	Total load, w_{ts} (lb/ft)	M_{ts} (K-in.)
1	91.79	4.0	3.10	7.10	98.9	76.8
2	86.29	4.0	3.14	7.14	93.4	72.5
3	81.84	4.0	3.10	7.10	88.9	69.0
4	186.4	4.0	5.05	9.05	195.4	151.7
5	189.1	4.0	5.01	9.01	198.1	153.8
6	184.5	4.0	4.91	8.91	193.4	150.2

B. Reduction Factor Data

Test no.	t (in.)	S_{eff} (in.)	F_y (ksi)	$M_{nt} = S_{eff}F_y$ (K-in.)	M_{ts} (K-in.)	R_t (%)
1	0.059	1.88	60.0	112.8	76.8	68.1
2	0.059	1.90	59.2	112.5	72.5	64.5
3	0.060	1.92	57.3	110.0	69.0	62.8
Average				111.8		65.1
4	0.097	3.38	68.4	231.2	151.7	65.6
5	0.096	3.38	67.1	226.8	153.8	67.8
6	0.097	3.30	66.5	219.4	150.2	68.5
Average				225.8		67.3

σ_{max} = one standard deviation of the modification factors of the thickest purlins tested = 0.0148

σ_{min} = one standard deviation of the modification factors of the thinnest purlins tested = 0.0271

$R_{t\ max} = 0.673 - 0.0148 = 0.658$

$R_{t\ min} = 0.651 - 0.0271 = 0.624$

$\overline{M}_{nt\ max} = 225.8$ kip-in.

$\overline{M}_{nt\ min} = 111.8$ kip-in.

C. Reduction Factor Relation
Equation (10.4):

$$R = \left(\frac{0.658 - 0.624}{225.8 - 111.8}\right)(M_n - 111.8) + 0.624 \leq 1.0$$

$$= \frac{0.298(M_n - 111.8)}{1000} + 0.624$$

The variation of R with purlin strength for the standing seam roof system tested is shown in Figure 10.20. (Figure 10.20 shows the R-value line sloping upward to the right. For some standing seam roof systems, the line will slope downward to the right.)

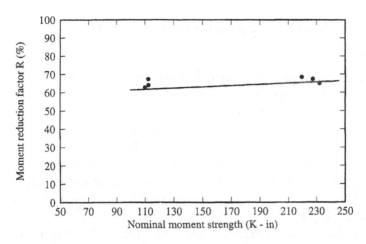

FIGURE 10.20   Reduction factor vs. nominal moment strength.

D. Application

For a purlin having the same depth, flange width, edge stiffener slope, and material specification as those used in the above example and with a nominal flexural strength $M_n = S_e F_y = 135$ in.-kips, the reduction factor is

$$R = \frac{0.298(135 - 111.77)}{1000} + 0.624 = 0.631$$

The positive moment design strength:

1. LRFD

$$\phi_b = 0.90$$
$$\phi M_n = \phi_b R S_e F_y = 0.90 \times 0.631 \times 135 = 76.7 \text{ kip-in.}$$

2. ASD

$$\Omega_b = 1.67$$
$$M = \frac{R S_e F_y}{\Omega_b} = \frac{0.631 \times 135}{1.67} = 51.0 \text{ kip-in.}$$

10.5.2 Continuous Lapped Z-Section Purlin

Problem

Determine the maximum uplift and gravity service loads for a three-span continuous purlin line using the Z-section purlins (8Z060) shown in Figure 4.14. The spans are 25 ft with interior lap lengths of 5 ft (3-ft end spans and 2-ft interior span) as shown in Figure 10.21. The purlin lines are 5 ft on center, and the roof sheathing is screw-fastened to the top flange of the purlins. The weight of the sheathing, insulation, and purlin is equivalent to 13.5 pounds per linear foot (lb/ft).

The following section, equation, and table numbers refer to those in the AISI Specification.

$$F_y = 55 \text{ ksi} \quad E = 29{,}500 \text{ ksi} \quad H = 8 \text{ in.} \quad y_{cg} = 4 \text{ in.}$$

Solution

A. Full Section Properties
    Area (A), second moment of area about centroidal axis perpendicular to web (I_x), second moment of area about

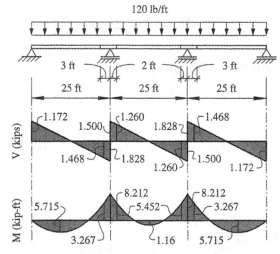

FIGURE 10.21 Three-span continuous lapped Z-purlin.

centroidal axis parallel to the web (I_y), and section modulus of compression flange (wide) for full section (S_f) accounting for rounded corners (from THIN-WALL (Ref. 11.3)):

$$A = 0.8876 \text{ in.}^2 \quad I_x = 8.858 \text{ in.}^4 \quad I_y = 1.714 \text{ in.}^4$$

$$I_{yc} = \frac{I_y}{2} = 0.857 \text{ in.}^4$$

$$S_f = \frac{I_x}{y_{cg}} = 2.215 \text{ in.}^3$$

B. **Design Load on Continuous Lapped Purlin under Uplift Loading**

1. **Strength for Bending Alone Braced by Sheathing**
   Section C3.1.3 Beams Having One Flange Through-fastened to Sheathing
   a. Uplift Loading: Continuous Lapped Z-Section
   LRFD

$$R = 0.70$$

$$S_e = 1.618 \text{ in.}^3 \qquad \text{(see Example 4.6.4)}$$

$$M_n = RS_eF_y = 62.29 \text{ kip-in.} = 5.191 \text{ kip-ft}$$

$$\phi_b = 0.90 \quad \text{(LRFD)}$$

$$\phi_b M_n = 4.672 \text{ kip-ft} \qquad \text{(Eq. C3.1.4-1)}$$

LRFD design load per unit length (see moment diagram in Figure 10.21 $w = 120 \, \text{lb/ft}$, where maximum moment $= 5.715 \, \text{kip-ft}$).

$$w_u \text{ (LRFD)} = \frac{\phi_b M_n \times 120}{5.715} = 98.1 \text{ lb/ft} \quad \text{(LRFD)}$$

ASD

$$\Omega_b = 1.67 \quad \text{(ASD)}$$

$$\frac{M_n}{\Omega_b} = 3.108 \text{ kip-ft}$$

ASD design load per unit length (see moment diagram in Figure 10.21 at $w = 120\,\text{lb/ft}$, where maximum moment $= 5.715\,\text{kip-ft}$):

$$w\ (\text{ASD}) = \frac{M_n}{\Omega_b} \times \frac{120}{5.715} = 65.3\ \text{lb/ft} \quad (\text{ASD})$$

2. **Strength for Shear Alone**
    Section C3.2 Strength for Shear Only

$$k_v = 5.34 \text{ for an unreinforced web}$$

$$h = H - 2(R + t) = 8 - 2(0.20 + 0.06)$$

$$= 7.48 \text{ in.}$$

$$\frac{h}{t} = \frac{7.48}{0.060} = 124.7$$

$$\sqrt{\frac{Ek_v}{F_y}} = \sqrt{\frac{29{,}500 \times 5.34}{55}} = 53.51$$

Since

$$\frac{h}{t} > 1.415\sqrt{\frac{Ek_v}{F_y}} = 75.73$$

then

$$V_n = \frac{0.905Ek_v t^3}{h} = 4.12 \text{ kips} \qquad ((\text{Eq. C3.2-3})$$

$$\phi_v = 0.90 \quad (\text{LRFD})$$

$$\Omega_v = 1.67 \quad (\text{ASD})$$

Maximum V at end of lap in end span $= 1.468$ kips when $w = 120\,\text{lb/ft}$

LRFD
Assuming that bending alone controls
w_u (LRFD) = 98.1 lb/ft (B.1), then

$$V_u = \frac{98.1}{120} \times 1.468 = 1.20 \text{ kips}$$

$$\frac{V_u}{\phi_v V_n} = \frac{1.20}{0.90 \times 4.12} = 0.324 < 1.0$$

Hence, shear alone is satisfied for LRFD under uplift loading.

ASD
Assuming that bending alone controls
w (ASD) = 65.3 lb/ft (B.1), then

$$V = \frac{65.3}{120} \times 1.468 = 0.80 \text{ kips}$$

$$\frac{V}{V_n/\Omega_v} = \frac{0.80}{4.12/1.67} = 0.324 < 1.0$$

Hence, shear alone is satisfied for ASD under uplift loading.

3. Strength for Bending at End of Lap
   Section C3.1.1   Nominal Section Strength

$$S_e = 1.618 \text{ in.}^3 \quad (\text{at } F_y)$$

$$M_{nxo} = S_e F_y \qquad \qquad (\text{see Example 4.6.4})$$

$$= 1.618 \times 55 \qquad \qquad (\text{Eq. C3.1.1-1})$$

$$= 89.0 \text{ kip-in.} - 7.42 \text{ kip-ft}$$

$$\phi_b = 0.95 \quad (\text{LRFD})$$

$$\Omega_b = 1.67 \quad (ASD)$$

Maximum M at end of lap in interior span — 5.452 kip-ft when w = 120 lb/ft.

LRFD

Again assuming that bending alone controls with w_u (LRFD) $= 98.1$ lb/ft (B.1), then

$$M_u = \frac{98.1}{120} \times 5.452 = 4.457 \text{ kip-ft}$$

$$\frac{M_u}{\phi_b M_{nxo}} = \frac{4.457}{0.95 \times 7.42} = 0.632 < 1.0$$

ASD

Again assuming that bending alone controls with w (ASD) $= 65.3$ lb/ft (B.1), then

$$M = \frac{65.3}{120} \times 5.452 = 2.97 \text{ kip-ft}$$

$$\frac{M}{M_{nxo}/Q_b} = \frac{2.97}{7.42/1.67} = 0.668 < 1.0$$

4. Strength for Combined Bending and Shear
   Section C3.5 Combined Bending and Shear
   LRFD
   For M_u and V_u based on bending alone (B.1) with w_u (LRFD) $= 98.1$ lb/ft. Maximum V at end of lap in interior span $= 1.260$ kips when $w = 120$ lb/ft. Hence

$$\frac{V_u}{\phi_v V_n} = \frac{98.1/120 \times 1.260}{0.90 \times 4.12} = 0.278$$

$$\left(\frac{M_u}{\phi_b M_{nxo}}\right)^2 + \left(\frac{V_u}{\phi_v V_n}\right)^2 = (0.632)^2 + (0.278)^2$$

$$= 0.477 < 1.0$$

Hence, combined bending and shear are satisfied for LRFD at end of lap in interior span. Checking end of lap in end span produces a lower (safer) result due to significantly lower moment.

ASD

For M and V based on bending alone (B.1) with w (ASD) $= 65.3\,\text{lb/ft}$:

$$\left(\frac{M}{M_{nxo}/\Omega_b}\right)^2 + \left(\frac{V}{V_n/\Omega_v}\right)^2 = (0.668)^2 + (0.278)^2$$

$$= 0.524 < 1.0$$

Hence, combined bending and shear are satisfied for ASD at end of lap in interior span. Checking end of lap in end span produces a lower result due to significantly lower moment.

C. Design Load on Continuous Lapped Purlin under Gravity Loading

    1. Strength for Bending with Lateral Buckling
       Section C3.1.2 Lateral Buckling Strength
          The Center for Cold-Formed Steel Structures Bulletin (Vol. 1, No. 2, August 1992) suggests using a distance from inflection point to end of lap, as unbraced length for gravity load with $C_b = 1.75$. Check interior span since it is more highly loaded in negative moment region.

          $L = $ distance from end of lap to inflection point as shown in Figure 10.21
            $= 73.1\,\text{in.}$ based on statics using Figure 10.21 with zero moment at inflection point

        $C_b = 1.75$

$$M_e = \frac{\pi^2 E C_b d I_{yc}}{2L^2} = 326.8 \text{ kip-in.} = 27.2 \text{ kip-ft}$$

$$\text{(Eq. C3.1.2-16)}$$

Critical Moment (M_c)

$$M_y = S_f F_y = 121.8 \text{ kip-in.} = 10.15 \text{ kip-ft}$$

$$\text{(Eq. C3.1.2-5)}$$

Since

$$M_e = 27.2 \text{ kip-ft} < 2.78 M_y = 28.2 \text{ kip-ft}$$

$$M_c = \frac{10}{9} M_y \left(1 - \frac{10 M_y}{36 M_e}\right) = 121.20 \text{ kip-in.}$$

(Eq. C3.1.2-3)

$$F_c = \frac{M_c}{S_f} = 54.73 \text{ ksi}$$

S_c is the effective section modulus at the stress, $F_c = 54.73$ ksi. Using the method in Example 4.6.4, but with $f = F_c$, gives

$$S_c = 1.621 \text{ in.}^3$$

$$M_n = S_c F_c = 88.99 \text{ kip-in.} = 7.416 \text{ kip-ft}$$

(Eq. C3.1.2-1)

LRFD

$$\phi_b = 0.90 \quad \text{(LRFD)}$$
$$\phi_b M_n = 6.674 \text{ kip-ft}$$

LRFD design load per unit length (see moment diagram in Figure 10.21 at $w = 120$ lb/ft, where maximum moment in interior span equals 5.452 kip-ft):

$$w_u \text{ (LRFD)} = \frac{\phi_b M_n \times 120}{5.452} = 146.9 \text{ lb/ft}$$

ASD

$$\Omega_b = 1.67 \quad \text{(ASD)}$$

$$\frac{M_n}{\Omega_b} = 4.44 \text{ kip-ft}$$

ASD design load per unit length (see moment diagram in Figure 10.21 at $w = 120$ lb/ft, where maximum moment in interior span equals 5.452 kip-ft):

$$w \text{ (ASD)} = \frac{M_n}{\Omega_b} \times \frac{120}{5.452} = 97.74 \text{ lb/ft}$$

2. Strength for Shear Alone
   Section C3.2 Strength for Shear Only
   See B.2 since loading direction does not affect shear strength.

$$V_n = 4.12 \text{ kips}$$
$$\phi_v = 0.90 \quad \text{(LRFD)}$$
$$\Omega_v = 1.67 \quad \text{ASD}$$

LRFD
Assuming that bending alone controls with $w_u = 146.9$ lb/ft (C.1), then

$$V_u = \frac{146.9}{120} \times 1.468 = 1.797 \text{ kips}$$

$$\frac{V_u}{\phi_v V_n} = \frac{1.797}{0.90 \times 4.12} = 0.485 < 1.0$$

Hence, shear alone is satisfied for LRFD under gravity loading.

ASD
Assuming that bending alone controls with w (ASD) = 97.74 lb/ft (C.1), then

$$V = \frac{97.74}{120} \times 1.468 = 1.196 \text{ kips}$$

$$\frac{V}{V_n/\Omega_v} = \frac{1.196}{4.12/1.67} = 0.485 < 1.0$$

Hence, shear alone is satisfied for ASD under gravity loading.

3. Strength for Bending at End of Lap
   Section C3.1.1   Nominal Section Strength
   See B.3 since loading direction does not affect bending strength.

   $$M_{nxo} = 7.42 \text{ kip-ft}$$

   $$\phi_b = 0.95 \quad \text{(LRFD)}$$

   $$\Omega_b = 1.67 \quad \text{(ASD)}$$

   Maximum M at end of lap in interior span = 5.452 kip-ft when $w = 120\,\text{lb/ft}$.

   LRFD
   Assuming that bending alone controls with w_u (LRFD) = 146.9 lb/ft (C.l), then

   $$M_u = \frac{146.9}{120} \times 5.452 = 6.654 \text{ kip-ft}$$

   $$\frac{M_u}{\phi_v M_{nxo}} = \frac{6.654}{0.95 \times 7.42} = 0.944 < 1.0$$

   Hence, bending alone at end of lap is satisfied for LRFD under gravity loading.

   ASD
   Assuming that bending alone controls with w (ASD) = 97.74 lb/ft (C.1), then

   $$M = \frac{97.74}{120} \times 5.452 = 4.44 \text{ kip-ft}$$

   $$\frac{M}{M_{nxo}/\Omega_b} = \frac{4.40}{7.42/1.67} = 1.0 \leq 1.0$$

   Hence bending alone at end of lap is satisfied for ASD under gravity loading.

4. Strength for Combined Bending and Shear
   Section C3.5 Combined Bending and Shear

LRFD
For M_u and V_u based on bending alone (C.1) with w_u (LRFD) $= 146.9\,$lb/ft. Maximum V at end of lap in interior span $= 1.260$ kips when $w = 120\,$lb/ft. Hence

$$\frac{V_u}{\phi_v V_n} = \frac{146.9/120 \times 1.260}{0.90 \times 4.12}$$

$$= 0.415$$

$$\frac{M_u}{\phi_b M_{nxo}} = 0.944 \qquad \text{(from C.3)}$$

$$\left(\frac{M_u}{\phi_b M_{nxo}}\right)^2 + \left(\frac{V_u}{\phi_v V_n}\right)^2 = (0.944)^2 + (0.415)^2$$

$$= 1.064 > 1.00$$

Hence, combined bending and shear are *not satisfied* for LRFD. Revise load down to

$$w_u \text{ (LRFD)} = \frac{146.9}{\sqrt{1.064}} = 142 \text{ lb/ft}$$

Checking end of lap in end span produces a lower (safer) result due to significantly lower moment.

ASD
For M and V based on bending alone (C.1) with $w = 97.74\,$lb/ft. Maximum V at end of lap in

341

interior span $= 1.260$ kips when $w = 120\,\text{lb/ft}$. Hence

$$\frac{V}{V_n/\Omega_v} = \frac{97.74/120 \times 1.260}{4.12/1.67}$$

$$= 0.415$$

$$\frac{M}{M_{nxo}/\Omega_b} = 1.00 \qquad \text{(from C.3)}$$

$$\left(\frac{M}{M_{nxo}/\Omega_b}\right)^2 + \left(\frac{V}{V_n/\Omega_v}\right)^2 = (1.00)^2 + (0.415)^2$$

$$= 1.172 > 1.00$$

Hence combined bending and shear is *not satisfied* for ASD. Revise load down to

$$w\,(\text{ASD}) = \frac{97.34}{\sqrt{1.172}} = 90.3 \text{ lb/ft}$$

Checking end of lap in end span produces a lower (safer) result due to significantly lower moment.

5. Web Crippling
The web crippling strength at the end and interior supports must be checked according to Section C3.4 of the AISI Specification if the purlins are supported in bearing by the building rafter flanges. The method set out in Example 6.8.1 should be used. If the purlins are web-bolted to antiroll devices or web shear plates, bearing and tearout at the web bolts, as well as bolt shear rupture, must be checked.

D. Service Loads

The following calculations assume that web crippling or bearing and tearout is satisfied.

1. LRFD

    a. Uplift Loading

       Using load combination 6 in *Section A6.1.2, Load Factors and Load Combinations*, $(0.9D - 1.3W)$, the controlling design load 98.1 lb/ft (B.1), and applying Exception 3 "For wind load on individual purlins, girts, wall panels and roof decks, multiply the load factor for W by 0.9."

$$0.9 \times 13.5 - (1.3 \times 0.9)w_w = -98.1 \text{ lb/ft}$$

$$w_w = 94.2 \text{ lb/ft}$$

       which is equivalent to a service uplift load of 18.1 psf for a 5-ft purlin spacing.

    b. Gravity Loading

       Using load combination 2 $(1.2D + 1.6S)$ and the controlling design load 142 lb/ft (C.4) give

$$1.2 \times 13.5 + 1.6w_s = 142 \text{ lb/ft}$$

$$w_s = 78.6 \text{ lb/ft}$$

       which is equivalent to a service snow load of 15.7 psf for a 5-ft purlin spacing.

2. ASD

    a. Uplift Loading

       Using load combination 3 in *Section A5.1.2, Load Combinations* $(D + W)$ and the controlling design load 65.3 lb/ft (B.1) give

$$13.5 - w_w = 65.3 \text{ lb/ft}$$

$$w_w = 78.8 \text{ lb/ft}$$

which is equivalent to a service uplift load of 15.8 psf for a 5-ft purlin spacing.

b. Gravity Loading

Using load combination $2\,(D+S)$ and the controlling design load 90.3 lb/ft (C.4) give

$$13.5 + w_s = 90.3 \text{ lb/ft}$$
$$w_s = 78.6 \text{ lb/ft}$$

which is equivalent to a service snow load of 15.4 psf for a 5-ft purlin spacing.

Problem

Determine the gravity service load for the conditions of the previous problem except with standing seam roof sheeting.

Solution

From the base test method, the R-factor relationship for the standing seam system is

$$R = \frac{0.350(M_n - 70.6)}{1000} + 0.691$$

with

$$M_n = S_e F_y = 1.618 \times 55 = 89.0 \text{ kip-in.}$$

$$R = \frac{0.350(89.0 - 70.6)}{1000} + 0.691 = 0.697$$

LRFD

Positive moment design strength

$$\phi_b = 0.90$$
$$\phi_b M_n = \phi_b R S_e F_y = 0.90 \times 0.697 \times 89.0 = 55.8 \text{ kip-in.}$$
$$= 4.65 \text{ kip-ft}$$

LRFD design load per unit length (see moment diagram Figure 10.21 at 120 lb/ft, where maximum positive moment $= 5.715$ kip-ft):

$$w_u = \frac{\phi_b M_n \times 120}{5.715} = 97.6 \, \text{lb/ft}$$

Since 97.6 lb/ft is less than 142 lb/ft determined in the above problem for through-fastened sheeting, the controlling factored gravity loading is this value. Using load combination 6 in *Section A6.1.2, Load Factors and Load Combinations* $(1.2D + 1.6W)$ gives

$$1.2 \times 13.5 + 1.6 w_s = 97.6 \, \text{lb/ft}$$
$$w_s = 50.9 \, \text{lb/ft}$$

which is equivalent to a service snow load of 10.2 psf for a 5-ft purlin spacing.

ASD
    Positive moment design strength:

$$\Omega_b = 1.67$$

$$\frac{M_n}{\Omega_b} = \frac{RS_e F_y}{\Omega_b} = \frac{0.697 \times 89.0}{1.67} = 37.1 \, \text{kip-in.}$$

$$= 3.10 \, \text{kip-ft}$$

ASD design load per unit length (see moment diagram (Figure 10.21 at 120 lb/ft, where maximum positive moment $= 5.715$ kip-ft):

$$w = \frac{M_n/\Omega_b \times 120}{5.715} = 65.1 \, \text{lb/ft}$$

Since 65.1 lb/ft is less than 90.3 lb/ft determined in the above problem for through-fastened sheeting, the controlling service gravity loading is this value. Using load combination 3 in *Section A5.1.2, Load Combinations* $(D + W)$:

$$13.5 + w_s = 65.1\,\text{lb/ft}$$
$$w_s = 51.6\,\text{lb/ft}$$

which is equivalent to a service snow load of 10.3 psf for a 5-ft purlin spacing.

10.5.3 Anchorage Force Calculations

Problem

Determine anchorage forces for a three-span continuous system having six parallel purlin lines with support restraints. The purlin section is 8Z060, the span length is 25 ft, and purlin lines are spaced 5 ft apart. The roof slope is 2 : 12 (9.46°), and the roof sheathing shear stiffness is 3500 lb/in. Uniform gravity loads of 2.7 psf dead load and 15 psf live load are applied to the system. Use the method in Ref. 10.19 as described in Section 10.4.1 with $\delta b = b/3$.

$$d = 8.0\,\text{in.} \quad b = 2.5\,\text{in.} \quad t = 0.060\,\text{in.} \quad I_x = 8.15\,\text{in.}^4$$
$$I_{xy} = 2.30\,\text{in.}^4$$

From Table 10.1, for a multiple-span system with support anchorage only:

Exterior restraints $C_1 = 0.50$ $C_2 = 5.9$ $C_3 = 0.35$
Interior restraints $C_1 = 1.00$ $C_2 = 9.2$ $C_3 = 0.45$

Solution

A. Anchorage Force per Purlin

$$P_0 = \left[\left(\frac{I_{xy}}{2I_x} + \frac{\delta b}{d}\right)\cos\theta - \sin\theta\right]W_p \qquad \text{(Eq. 10.12)}$$

$$= \left[\left(\frac{2.30}{2(8.15)} + \frac{2.5}{3(8.0)}\right)\cos 9.46° - \sin 9.46°\right]W_p$$

$$= 0.07758 W_p \text{ lb}$$

B. Anchorage Force at Restraint

$$n_p^* = \min\{n_p, n_{p(\max)}\}$$

$$n_{p(\max)} = 0.5 + \frac{d}{2C_2 t} \qquad \text{(Eq.10.15)}$$

For the exterior restraints:

$$n_{p(\max)} = 0.5 + \frac{8.0}{2(5.9)(0.060)} = 11.80 > 6 \rightarrow n_p^* = 6$$

$$\alpha = 1 - C_2\left(\frac{t}{d}\right)(n_p^* - 1) \qquad \text{(Eq. 10.14)}$$

$$= 1 - 5.9\left(\frac{0.060}{8.0}\right)(6-1) = 0.7788$$

For the interior restraints:

$$n_{p(\max)} = 0.5 + \frac{8.0}{2(9.2)(0.060)} = 7.746 > 6 \rightarrow n_p^* = 6$$

$$\alpha = 1 - 9.2\left(\frac{0.060}{8.0}\right)(6-1) = 0.6550$$

Roof panel shear stiffness modifier:

$$\gamma = C_3 \log\left(\frac{G'}{2500}\right) \qquad \text{(Eq. 10.17)}$$

For the exterior restraints:

$$\gamma = 0.35 \log\left(\frac{3500}{2500}\right) = 0.05114$$

For the interior restraints:

$$\gamma = 0.45 \log\left(\frac{3500}{2500}\right) = 0.06576$$

Support anchorage force:

$$P_L = P_0 C_1(n_p^* \alpha + n_p \gamma) \qquad \text{(Eq. 10.10)}$$

Exterior anchorage force:

$$P_L = (0.07758 W_p)(0.50)[(6)(0.7788) + (6)(0.05114)]$$
$$= 0.1931 W_p \text{ lb} \quad \text{(tension)}$$

Interior anchorage force:

$$P_L = (0.07758 W_p)(1.00)[(6.0)(0.6550)$$
$$+ (6)(0.06576)]$$
$$= 0.3354 W_p \text{ lb} \quad \text{(tension)}$$

C. Design Anchorage Forces
    LRFD
    Using load combination 2 in *Section A6.1.2, Load Factors and Load Combinations* $(1.2D + 1.6S)$:

$$w_u = 1.2 \times 2.7 + 1.6 \times 15 = 27.24 \text{ psf}$$

The uniform load is evenly distributed to all purlin lines. The total average load on each purlin in a bay is then

$$W_{pu} = \frac{27.24 \times 5 \times 25 \times (6 - 1)}{6} = 2838 \text{ lb}$$

Exterior anchorage force:

$$P_{Lu} = 0.1931 \times 2838 = 548.0 \text{ lb}$$

Interior anchorage force:

$$P_{Lu} = 0.3354 \times 2838 = 951.8 \text{ lb}$$

ASD
Using load combination 3 in *Section A5.1.2, Load Combinations* $(D + W)$:

$$w = 2.7 + 15 = 17.7 \text{ psf}$$

The uniform load is evenly distributed to all purlin lines. The total average load on each purlin in a bay is then

$$W_{pu} = \frac{17.7 \times 5 \times 25 \times (6 - 1)}{6} = 1844 \text{ lb}$$

Exterior anchorage force:

$$P_L = 0.1931 \times 1844 = 356.1 \text{ lb}$$

Interior anchorage force:

$$P_L = 0.3354 \times 1844 = 618.5 \text{ lb}$$

The required anchorage forces are shown graphically in Figure 10.22.

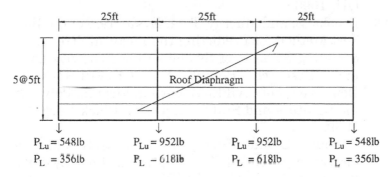

FIGURE 10.22 Anchorage forces for example.

REFERENCES

10.1 American Iron and Steel Institute, A Guide for Designing with Standing Seam Roof Panels, Design Guide 97-1, American Iron and Steel Institute, Washington, DC, 1997.

10.2 Bryant, M. R. and Murray, T. M., Investigation of Inflection Points as Brace Points in Multi-span Purlin Roof Systems, Research Report No. CE/VPI-ST-00/11, Department of Civil and Environmental Engineering, Virginia Polytechnic Institute and State University, Blacksburg, VA, 2000.

10.3 LaBoube, R. A., Golovin, M., Montague, D. J., Perry, D. C. and Wilson, L. L., Behavior of Continuous Span Purlin System, Proceedings of the Ninth International Specialty Conference on Cold-Formed Structures, University of Missouri-Rolla, Rolla, MO, 1988.

10.4 LaBoube, R. A., Roof Panel to Purlin Connection: Rotational Restraint Factor, Proceedings of the ISABSE Colloquium on Thin-Walled Structures in Buildings, Stockholm, Sweden, 1986.

10.5 Fisher, J. M., Uplift Capacity of Simple Span Cee and Zee Members with Through-Fastened Roof Panels, Final Report MBMA 95-01, Metal Building Manufacturers Association, Cleveland, OH, 1996.

10.6 Brooks, S. and Murray, T. M. Evaluation of the Base Test Method for Predicting the Flexural Strength of Standing Seam Roof Systems under Gravity Loading', Research Report No. CE/VPI-ST-89/07, Department of Civil Engineering, Virginia Polytechnic Institute and State University, Blacksburg, VA, 1989.

10.7 Rayburn, L. and Murray, T. M., Base Test Method for Gravity Loaded Standing Seam Roof Systems, Research Report CE/VPI-ST-90/07, Department of

Civil Engineering, Virginia Polytechnic Institute and State University, Blacksburg, VA, 1990.

10.8   Anderson, B. B. and Murray, T. M., Base Test Method for Standing Seam Roof Systems Subject to Uplift Loading, Research Report CE/VPI-ST-90/06, Department of Civil Engineering, Virginia Polytechnic Institute and State University, Blacksburg, VA, 1990.

10.9   Pugh, A. D. and Murray, T. M., Base Test Method for Standing Seam Roof Systems Subject to Uplift Loading—Phase II, Research Report CE/VPI-ST-91/17, Department of Civil Engineering, Virginia Polytechnic Institute and State University, Blacksburg, VA, 1991.

10.10   Mills, J. F. and Murray, T. M., Base Test Method for Standing Seam Roof Systems Subject to Uplift Loading—Phase III, Research Report CE/VPI-ST-92/09, Department of Civil Engineering, Virginia Polytechnic Institute and State University, Blacksburg, VA, 1992.

10.11   Ellifritt, D., Sputo, T. and Haynes, J., Flexural Capacity of Discretely Braced C's and Z's, Proceedings of the Eleventh International Specialty Conference on Cold-Formed Structures, University of Missouri-Rolla, Rolla, MO, 1992.

10.12   Murray, T. M. and Elhouar, S., North American approach to the design of continuous Z- and C-purlins for gravity loading with experimental verification, Engineering Structures, Vol. 16, No. 5, 1994, pp. 337–341.

10.13   Yura, J. A., Bracing for Stability—State-of-Art, University of Texas-Austin, 1999.

10.14   American Institute of Steel Construction, Load and Resistance Design Specification for Structural Steel Buildings, American Institute of Steel Construction, Chicago, IL, 1993.

10.15   Center for Cold-Formed Steel Structures, Bulletin Vol. 1, No. 2, August 1992.

10.16   Zetlin, L. and Winter, G., Unsymmetrical bending of beams with and without lateral bracing, Proceedings of the American Society of Civil Engineers, Vol. 81, 1955.

10.17   Elhouar, S. and Murray, T. M., Stability Requirements of Z-Purlin Supported Conventional Metal Building Roof Systems, Annual Technical Session Proceedings, Structural Stability Research Council, Bethlehem, PA, 1985.

10.18   Seshappa, V. and Murray, T. M., Study of Thin-Walled Metal Building Roof Systems Using Scale Models, Proceedings of the IABSE Colloquium on Thin-Walled Metal Structures in Buildings, IABSE, Stockholm, Sweden, 1986.

10.19   Neubert, M. C. and Murray, T. M., Estimation of Restraint Forces for Z-Purlins Roofs under Gravity Load, Proceedings of the Fifteenth International Specialty Conference on Cold-Formed Structures, University of Missouri-Rolla, Rolla, MO, 2000.

11

Steel Storage Racking

11.1 INTRODUCTION

Steel racks were introduced in Section 1.5, and typical components and configurations are drawn in Figures 1.4 and 1.5. A large proportion of steel storage rack systems are manufactured from cold-formed steel sections, so the design procedures set out in the AISI Specification and this book are relevant to the design of steel storage racks for supporting storage pallets. However, several aspects of the structural design of steel storage racks are not adequately covered by the AISI Specification and are described in the Rack Manufacturers Institute (RMI) Specification (Ref. 11.1). These aspects include the following:

(a) Loads specific to rack structures covered in Section 2, Loading
(b) Design procedures as described in Section 3
(c) Design of steel elements and members accounting for perforations (slots) as described in Section 4

(d) Determination of bending moments, reactions, shear forces, and deflections of beams accounting for partial end fixity, as given in Section 5

(e) Upright frame design, including effective length factors and stability of truss-braced upright frames as presented in Section 6

(f) Connections and bearing plate design requirements as presented in Section 7

(g) Special rack design provisions described in Section 8

(h) Test methods, including stub column tests, pallet beam tests, pallet beam to column connection tests, and upright frame tests as described in Section 9

The 1990 edition of the RMI Specification (Ref. 11.2) was developed in allowable stress format to be used in conjunction with the 1986 edition of the AISI Specification in allowable stress format. However, the most recent 1997 edition of the RMI Specification (Ref. 11.1) has been written to allow both ASD and LRFD. In addition, it is written to align with the 1996 edition of the AISI Specification (Ref. 1.2).

The areas where significant changes to the 1990 RMI Specification (Ref. 11.2) have been made in the 1997 RMI Specification (Ref. 11.1) are

(a) Load combinations and load factors for the LRFD method are specified in Section 2.2 of the 1997 RMI Specification and are discussed in Section 11.2 of this book.

(b) Design curves for beams and columns with perforations have been developed and are specified in Sections 4.2.2 and 4.2.3, respectively, of the 1997 RMI Specification. They are discussed in Section 11.4 of this book and have been used in the design example in Section 11.6.

(c) Earthquake forces are specified in Section 2.7 of the 1997 RMI Specification.

11.2 LOADS

Steel storage racking must be designed for appropriate combinations of dead loads, live loads (other than pallet and product storage), vertical impact loads, pallet and product storage loads, wind loads, snow loads, rain loads, or seismic loads. The major loading on racks is the gravity load resulting from unit (pallet) loads taken in conjunction with the dead load of the structure. In practice, since racks are usually contained within buildings and wind loads are not applicable, the major component of horizontal force is that resulting from the gravity loads, assuming an out-of-plumb of the upright frames forming the structure. In some installations earthquake forces may also need to be considered and may be greater than the horizontal force caused by gravity loads, assuming an out-of-plumb.

The horizontal forces resulting from out-of-plumb are specified in Section 2.5.1 of the RMI Specification and are shown in Figure 11.1, where

$$\theta = 0.015 \text{ rad} \tag{11.1}$$

The vertical loads on the frames with initial out-of plumb shown on the left- hand diagrams in Figures 11.1a and b are statically equivalent to horizontal and vertical loads on frames with no initial out-of-plumb with the magnitude of the forces as shown on the right-hand diagrams in Figures 11.1a and b. These are the loads which should be used in design. They should be applied separately, not simultaneously, in each of the two principal directions of the rack.

The Commentary to the RMI Specification suggests that when many columns are installed in a row and interconnected, the $P\Delta$ force in Figure 11.1b was balanced out. Further, the Commentary states that thousands of storage rack systems have been designed and installed without $P\Delta$ forces and have performed well. It is left to the reader to determine whether to heed this comment or use Eq. (11.1) with the $P\Delta$ forces in Figure 11.1.

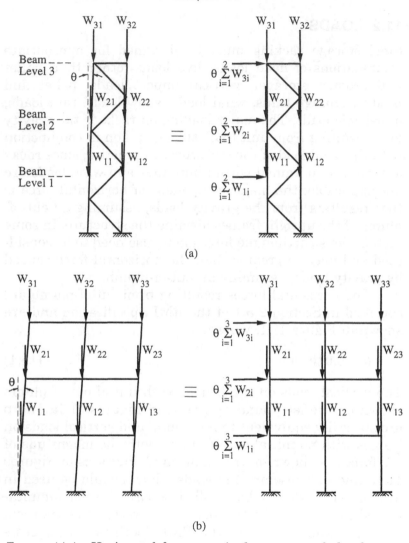

FIGURE 11.1 Horizontal forces equivalent to out-of-plumb gravity loads: (a) upright frames (cross-aisle); (b) plane of beams (down-aisle).

The 1997 RMI Specification provides loading combinations for storage rack systems. They are an extension of those in the 1996 AISI Specification (Ref. 1.2). The ASD combinations are given in Section 2.1, and the LRFD combinations are given in Section 2.2.

The additional loads to be considered in the design of storage racking are

PL = Maximum load from pallets or product stored on the racks

Imp = Impact loading on a shelf

$PLapp$ = That portion of pallet or product load which is used to compute seismic base shear

For ASD, PL is added to

Gravity Load Critical (Eq. 2, AISI Section A5.1.2) and Gravity Plus Wind/Seismic Critical (Eq. 4, AISI Section A5.1.2).

$PLapp$ is added to Uplift Critical (Eq. 3, AISI Section A5.1.2).
An additional combination of 5, $DL + LL + 0.5(SL \text{ or } RL) + 0.88PL + Imp$, is included for shelf plus Impact Critical in the RMI Specification.

For LRFD, factored values of PL are added to

Dead load (Eq. 1, AISI Section A6.1.2), factor = 1.2

Live/product load (Eq. 2, AISI Section A6.1.2), factor = 1.4

Snow/rain load (Eq. 3, AISI Section A6.1.2), factor = 0.85

Wind load (Eq. 4, AISI Section A6.1.2), factor = 0.85

Seismic load (Eq. 5, AISI Section A6.1.2), factor = 0.85

Uplift load (Eq. 6, AISI Section A6.1.2), factor = 0.9

An additional combination of 7, $1.2DL + 1.6LL + 0.5(SL \text{ or } RL) + 1.4PL + 1.4Imp$, is included for the Product/Live/Impact for shelves and connections.

The vertical impact load is 25% of an individual unit load. Vertical impact loads equivalent to the unit load factored by 1.25 should be used in the design of individual beams and connectors.

11.3 METHODS OF STRUCTURAL ANALYSIS

11.3.1 Upright Frames

The RMI Specification (Ref. 11.1) states that computations for safe loads, stresses, deflections, and the like shall be made in accordance with conventional methods of structural design as specified in the latest edition of the AISI Specification for cold-formed steel components and structural systems and the latest edition of the AISC Specification for hot-rolled steel components and systems except as modified or supplemented by the RMI Specification. Where adequate methods of design calculations are not available, designs shall be based on test results.

In the first-order method of analysis, the structure is analyzed in its undeformed configuration. The first-order bending moments are amplified within the beam-column interaction equations (see Section 8.2), using the buckling load of the frame based on the appropriate effective lengths. In the case of down-aisle stability, a value of $K_x = L_x/L$ of 1.7 is suggested if a rational buckling analysis is not performed. However, it is recommended that the value of effective length be determined accurately, accounting for the joint flexibility and the restraint provided at the column bases. Detailed equations are included in the Commentary to the RMI Specification in Section 6.3.1.1 (racks not braced against side sway) for the determination of the G_A and G_B factors when calculating the effective lengths using the AISI Specification. Example 11.6 demonstrates these calculations.

The influence of the joint flexibility is best determined experimentally using the portal test described in Section

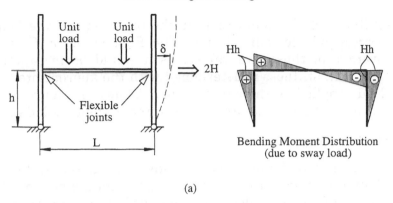

Bending Moment Distribution
(due to sway load)

(a)

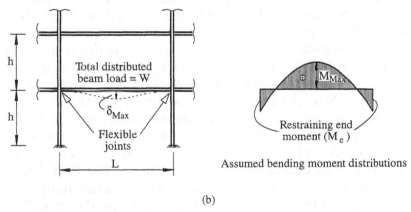

Assumed bending moment distributions

(b)

FIGURE 11.2 Influence of flexible joints: (a) portal test; (b) beam deflections and moments.

9.4.2 of the RMI Specification. The portal test involves determining the sway stiffness of a single frame as shown in Figure 11.2a. The method for evaluation of the test results is set out in the Commentary to the RMI Specification in Section 9.4.2.3 to determine the spring constant (F) relating the moment to the rotation of the flexible joints shown in Figure 11.2a. The advantage of the portal test in determining F is that the stiffness of the connection at one end of the beam corresponds to joint opening, and the

359

stiffness of the connection at the other end of the beam corresponds to joint closing. The final value of F is an average of these two values, as it would be in practice. The joint stiffness is determined with the appropriate unit loads applied on the pallet beam as shown in Figure 11.2a to ensure that the joints are measured in their loaded state.

11.3.2 Beams

In the design of pallet beams in steel storage racks, the end fixity produces a situation where the beam can be regarded as neither fixed-ended nor pin-ended. It is important to accurately account for the joint flexibility in the design of beams since the beam deflections and the bending moment distribution are highly dependent on the value of joint flexibility. If a very stiff joint is produced, the beam deflections will be reduced. However, the resulting restraining moment on the end of the beam may damage the joint. If a very flexible joint is used, the beam deflections may be excessive.

The deflected shape of a pallet beam and the assumed bending moment distribution are shown in Figure 11.2b. The restraining moment (M_e) is a function of the joint stiffness (F), which can be determined from either a cantilever test, as described in Section 9.4.1 of the RMI Specification, or the pallet beam in upright frame assembly test, as described in Section 9.3.2 of the RMI Specification. The resulting equations for the maximum bending moment ($M_{\max}$) and central deflection ($\delta_{\max}$) are given by Eqs. (11.2) and (11.3), which are based on the bending moment distribution in Figure 11.2b.

$$M_{\max} = \left(\frac{WL}{8}\right) r_m \qquad (11.2a)$$

$$r_m = 1 - \frac{2FL}{6EI_b + 3FL} \qquad (11.2b)$$

where W = total load on each beam
$\qquad L$ = span of beam

$F = $ joint spring constant

$E = $ modulus of elasticity

$I_b = $ beam moment of inertia about the bending axis and column, respectively

$$\delta_{\max} = \delta_{ss} r_d \tag{11.3a}$$

$$\delta_{ss} = \frac{5WL^3}{384EI_b} \tag{11.3b}$$

$$r_d = 1 - \frac{4FL}{5FL + 10EI_b} \tag{11.3c}$$

Section 5.2 of the RMI Commentary recommends that the maximum value of deflection at the center of the span not exceed 1/180 of the span measured with respect to the ends of the beam.

11.3.3 Stability of Truss-Braced Upright Frames

To prevent tall and narrow truss-braced upright frames of the type shown in Figure 11.1a from buckling in their own plane, elastic critical loads (P_{cr}) are specified in Section 6.4 of the 1997 RMI Specification. The elastic critical loads can be used to compute an equivalent effective length for use in the AISI Specification for cold-formed members or the AISC Specification for hot-rolled members.

11.4 EFFECTS OF PERFORATIONS (SLOTS)

The uprights of steel storage racks usually contain perforations (slots) where pallet beams and bracing members are connected into the uprights. These perforations can produce significant reductions in the bending and axial capacity of the upright sections. The design rules for flexural members use the elastic section modulus of the net section ($S_{x,\text{net min}}$). The design rules for axially loaded compression members use the minimum cross-sectional area ($A_{\text{net min}}$). Hence, it is necessary to determine these properties for the upright sections.

11.4.1 Section Modulus of Net Section

A perforated beam section in bending is shown in Figure 11.3a. The effect of the slot in the flanges is to reduce the second moment of area and hence the section modulus ($S_{x,\text{net min}}$) of the net section. $S_{x,\text{net min}}$ is the minimum value of the section modulus with the holes removed.

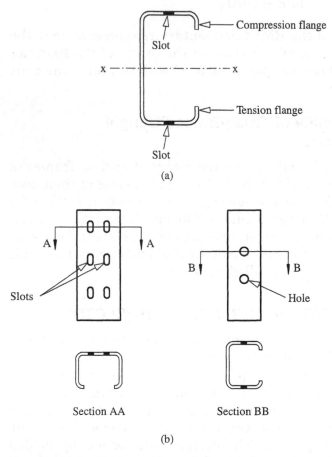

(a)

Section AA   Section BB

(b)

FIGURE 11.3   Perforated sections: (a) flexural member; (b) compression member.

362

A perforated compression member is shown in Figure 11.3b. The slots in the web are staggered so that they do not coincide with the holes in the flanges. Consequently, when a plane is passed through Section AA, a different net area will be determined from the case where a plane is passed through Section BB. The minimum net area ($A_{\text{net min}}$) is the lesser of these two values. If the slots and holes coincide, the minimum net area must have both slots and holes deducted from the total area.

11.4.2 Form Factor (Q)

The form factor (Q) allows for the effects of local buckling and postbuckling on the stub column strength of the section with perforations. The Q factor cannot be determined theoretically, so a test procedure is described in Section 9.2 of the RMI Specification. The value of Q is

$$Q = \frac{\text{Ultimate compressive strength of stub upright by test}}{F_y A_{\text{net min}}} \qquad (11.4)$$

where F_y = actual yield stress of the upright material. The value of Q allows for the interaction of local buckling with the perforations, but does not allow for the reduced area resulting from the perforations, since this has already been accounted for by using $A_{\text{net min}}$ in the design equations, as described in Section 11.5.

11.5 MEMBER DESIGN RULES

The member design rules for flexural and compression members specified in the 1996 AISI Specification have been modified in the 1997 RMI Specification to take account of the perforations. The values of $S_{x,\text{net min}}$ and $A_{\text{net min}}$ are used in conjunction with the design rules in the AISI Specification to produce moment and axial capacities.

11.5.1 Flexural Design Curves

The effective section modulus (S_e) with the extreme compression or tension fiber at F_y and specified in Section C3.1.1 of the AISI Specification for computing the nominal flexural strength (M_n) is modified in the 1997 RMI Specification to

$$S_e = \left(0.5 + \frac{Q}{2}\right)S_{x,\text{net min}} \tag{11.5}$$

The form factor Q is determined from a stub column test as described in Section 11.4.2 and in Section 9.2 of the 1997 RMI Specification.

The effective section modulus (S_c) at the critical stress $(F_c = M_c/Z_f)$ in the extreme compression fiber and specified in Section C3.1.2 of the AISI Specification for computing the nominal flexural strength (M_n) is modified in the 1997 RMI Specification to

$$S_c = \left(1 - \left(\frac{1-Q}{2}\right)\left(\frac{M_c/S_f}{F_y}\right)^Q\right)S_{x,\text{net min}} \tag{11.6}$$

In Eqs. (11.5) and (11.6), the effect of local buckling and the interaction of local buckling with perforations are accounted for by using the form factor Q derived from a stub column test. As M_c and F_c decrease, S_c approaches the section modulus of the net section $(S_{x,\text{net min}})$. As the value of $F_c = M_c/S_f$ approaches F_y, S_c approaches that given by Eq. (11.5), where the effect of Q is essentially half what it would be for a stub column since only one flange is in compression, as shown in Figure 11.3a. The resulting design curves for different values of Q are shown in Figure 11.4, where M_e is the elastic critical moment determined according to Section C3.1.2 of the AISI Specification.

11.5.2 Column Design Curves

The effective area (A_e) at the buckling stress (F_n) and specified in Section C4 of the AISI Specification for computing the nominal axial strength (P_n) is modified in the 1997

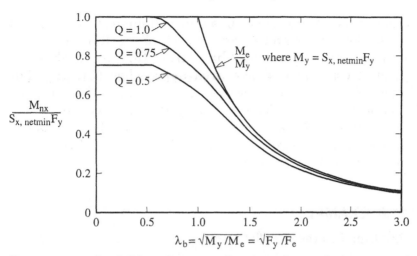

FIGURE 11.4   Rack Manufacturers Institute beam design curves.

RMI Specification (Ref. 11.1) to

$$A_e = \left(1 - (1 - Q)\left(\frac{F_n}{F_y}\right)^Q\right)A_{\text{net min}} \qquad (11.7)$$

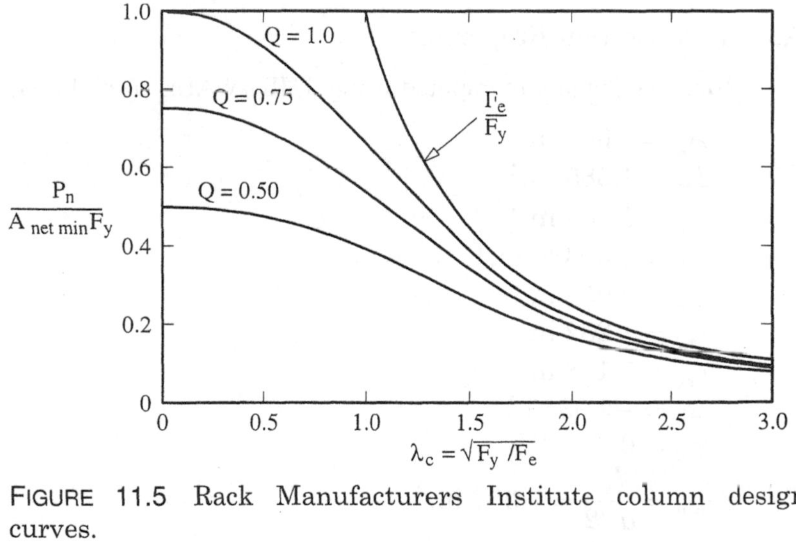

FIGURE 11.5   Rack Manufacturers Institute column design curves.

where Q is determined from a stub column test as described in Section 11.4.2 and in Section 9.2 of the RMI Specification. As the nominal buckling stress (F_n) approaches the yield stress, the effective area approaches $QA_{\text{net min}}$. For long columns with a low value of F_n, the effective area approaches $A_{\text{net min}}$. The resulting design curves for different values of Q are shown in Figure 11.5, where F_c is the elastic buckling stress determined according to Section C4 of the AISI Specification.

11.6 EXAMPLE

Problem: Unbraced Pallet Rack

Calculate the design load in the uprights for the three-level multibay rack in Figure 11.6b. The upright section is shown in Figure 11.6a. The design example is the same frame and section as in Problem 1, Part III of Ref. 11.2. Section numbers refer to the specification referenced in each section (**A** to **I**).

Solution

A. Gross Section Properties

Rounded Corners [computed using THIN-WALL (Ref. 11.3)]

$$A_g = 1.046 \text{ in.}^2$$
$$I_{xp} = 1.585 \text{ in.}^4$$
$$I_{yp} = 1.304 \text{ in.}^4$$
$$J = 0.00385 \text{ in.}^4$$
$$C_w = 3.952 \text{ in.}^6$$
$$r_{xg} = 1.232 \text{ in.}$$
$$r_{yg} = 1.118 \text{ in.}$$
$$x_0 = -2.935 \text{ in.}$$
$$y_0 = 0$$
$$S_{x\min} = \frac{I_{xp}}{a'/2} = 1.057 \text{ in.}^3$$

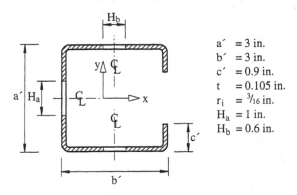

a´ = 3 in.
b´ = 3 in.
c´ = 0.9 in.
t = 0.105 in.
r_i = $^3/_{16}$ in.
H_a = 1 in.
H_b = 0.6 in.

Web and flange holes are staggered

(a)

L_{C1}= 48 in.

L_{C1}= 48 in.

L_{C2}= 60 in.

L_{long}= 42 in.

L_{short}= 4 in.

L_b = 100 in.

(b)

FIGURE 11.6 Unbraced storage rack example: (a) perforated lipped channel; (b) frame geometry.

367

RMI Specification Section 4.2.2: r_0, C_w should be based on the gross cross section, assuming square corners.

<u>Square Corners</u> [computed using THIN-WALL (Ref. 11.3)]

$A_g = 1.090$ in.^{2}

$I_{xp} = 1.683$ in.^{4}

$I_{yp} = 1.404$ in.^{4}

$J = 0.0040$ in.^{4}

$C_w = 4.213$ in.^{6}

$x_0 = -2.944$ in.

$r_0^2 = \dfrac{I_{xp} + I_{yp}}{A_g} + x_0^2 = 3.39$ in.^{2}

B.   Net Section Properties

<u>Rounded Corners</u> [computed using THIN-WALL (Ref. 11.3)]

$H_a = $ removed

$A_{yn} = 0.941$ in.^{2}

$I_{xp} = 1.576$ in.^{4}

$I_{yp} = 1.115$ in.^{4}

$r_{yn} = \sqrt{\dfrac{I_{yp}}{A_{yn}}} = 1.089$ in.

$H_b = $ removed

$A_{xn} = 0.920$ in.^{2}

$I_{xp} = 1.321$ in.^{4}

$I_{yp} = 1.296$ in.^{4}

$r_{xn} = \sqrt{\dfrac{I_{xp}}{A_{xn}}} = 1.198$ in.

C.   Effective Length Factors

RMI Commentary—Section 6.3.1.1: Racks Not Braced against Side Sway

$$\frac{I_f}{L_f} = \frac{bd^2}{1440} = \frac{3 \times 3^2}{1440} = 0.019 \text{ in.}^3$$

From Problem 1, Part III of Ref. 11.1:

$$I_b = 8.871 \text{ in.}^4$$

Measured Joint Stiffness

$$F = 10 \times 10^5 \text{ in. lb/rad}$$

$$\left(\frac{I_b}{L_b}\right)_{\text{red}} = \frac{I_b/L_b}{1 + 6EI_b/L_bF}$$

$$= \frac{8.871/100}{1 + \dfrac{6 \times 29500 \times 8.871}{100 \times 10 \times 10^5}}$$

$$= 0.00531 \text{ in.}^3$$

$$L_{c1} = 48 \text{ in.}, \qquad L_{c2} = 60 \text{ in.}$$

$$G_A = \frac{I_c(1/L_{c1} + 1/L_{c2})}{2(I_b/L_b)_{\text{red}}} = \frac{1.585(1/48 + 1/60)}{2 \times 0.00531}$$

$$= 5.60$$

$$G_B = \frac{I_c/L_{c2}}{I_f/L_f} = \frac{1.585/60}{0.019} = 1.409$$

$$G_A = 5.60, \qquad G_B = 1.409$$

$$K_x = 1.8 \qquad \text{(see AISC Specification Commentary Ref. 1.1)}$$

For this problem, it can be shown that the portion of the column between the floor and the first beam level governs. In general, all portions of columns should be checked.

RMI Specification—Section 6.3.2: Flexural Buckling in the Plane of the Upright Frame: Section 6.3.2.2

$$\frac{L_{\text{short}}}{L_{\text{long}}} = \frac{4}{42} = 0.095 < 0.15$$

Hence effective length factor (K_y) can be taken as

$$K_y = 1.0$$

RMI Specification—Section 6.3.3: Torsional Buckling: Section 6.3.3.2

$K_t = 0.8$

D. Torsional-Flexural Buckling Parameters

$$L_x = L_{C2} \qquad K_x L_x = 1.8 \times 60 = 108 \text{ in.}$$
$$L_y = L_{\text{long}} \qquad K_y L_y = 1.0 \times 42 = 42 \text{ in.}$$
$$L_t = L_{\text{long}} \qquad K_t L_t = 0.8 \times 42 = 33.6 \text{ in.}$$

AISI Specification—Section C4: Concentrically Loaded Compression Members

$$\sigma_{ex} = \frac{\pi^2 E}{(K_x L_x / r_{xg})^2} = \frac{\pi^2 \times 29500}{(108/1.232)^2} = 37.89 \text{ ksi}$$

$$\text{(Eq. C3.1.2-8)}$$

$$\sigma_{ey} = \frac{\pi^2 E}{(K_y L_y / r_{yg})^2} = \frac{\pi^2 \times 29500}{(42/1.118)^2} = 206.3 \text{ ksi}$$

$$\text{(Eq. C3.1.2-9)}$$

Using RMI Specification—Section 4.2.2, r_0 and C_w should be based on Gross Section assuming square corners:

$A = A_g = 1.046 \text{ in.}^2$ \qquad (full section round corners)

$r_0 = 3.39 \text{ in.}$ \qquad (full section square corners)

$C_w = 4.213 \text{ in.}^6$ \qquad (full section square corners)

$$\sigma_t = \frac{GJ}{Ar_0^2}\left(1 + \frac{\pi^2 EC_w}{GJ(K_t L_t)^2}\right) \qquad \text{(Eq. C3.1.2-10)}$$

$$= \frac{11346 \times 0.00385}{1.046 \times (3.39)^2}\left(1 + \frac{\pi^2 \times 29500 \times 4.213}{11346 \times 0.00385 \times (33.6)^2}\right)$$

$$= 94.1 \text{ ksi}$$

AISI Specification—Section C4.2

Using $x_0 = -2.944$ in., $r_0 = 3.39$ in.

$$\sigma_{tf0} = \frac{(\sigma_{ex} + \sigma_t) - \sqrt{(\sigma_x + \sigma_t)^2 - 4(1 - (x_0/r_0)^2)\sigma_{ex}\sigma_t}}{2(1 - (x_0/r_0)^2)}$$

$F_e = \sigma_{tf0} = 28.53$ ksi $< \sigma_{ey} = 206.3$ ksi   (Eq. C4.2-1)

Torsional-flexural buckling controls.

E.   Upright Column Concentric Strength

AISI   Specification—Section   C4.1:   Concentrically Loaded Compressive Members

$Q = 0.82$ from stub column test

$F_y = 45$ ksi

$F_e = 28.53$ ksi

$$\lambda_c = \sqrt{\frac{F_y}{F_e}} = \sqrt{\frac{45}{28.53}} = 1.256 < 1.5 \qquad \text{(Eq. C4-4)}$$

$F_n = (0.658^{\lambda_c^2})F_y = (0.658^{1.578}) \times 45 = 23.26$ ksi

(Eq. C4-2)

$A_{\text{net min}} = A_{xn} = 0.92$ in.^{2}

From RMI Specification—Section 4.2.3

$$A_e = \left(1 - (1 - Q)\left(\frac{F_n}{F_y}\right)^Q\right)A_{\text{net min}}$$

$$= \left(1 - (1 - 0.82)\left(\frac{23.26}{45}\right)^{0.82}\right)0.92 \text{ in.}^2$$

$= 0.824$ in.^{2}

$P_n = A_e F_n = 0.824 \times 23.26$ kips $= 19.15$ kips

F.   Flexural Strength for Bending of Upright about Plane of Symmetry (x-axis)

AISI Specification—Section C3.1.2
Singly symmetric section bent about the symmetry axis

$$M_e = C_b A r_0 \sqrt{\sigma_{ey}\sigma_t} \qquad\qquad \text{(Eq. C3.1.2-6)}$$

From section D,

$\sigma_{ey} = 206.3$ ksi

$\sigma_t = 94.1$ ksi

$A_g = 1.046$ in.^{2} (full section round corners)

$r_0 = 3.39$ in.   (full section square corners)

$C_b = 1.0$ (members subject to combined axial load
and bending moment)

$M_e = 1.0 \times 1.046 \times 3.39 \times \sqrt{206.3 \times 94.1} = 494.2$ kip-in.

$M_y = S_f F_y \qquad\qquad \text{(Eq. C3.1.2-5)}$

$S_f = S_{x\text{min}} = 1.057$ in.^{3}

$M_y = 1.057 \times 45 = 47.55$ kip-in.

Now, $M_e = 494.2 > 2.78\, M_y = 132.2$ kip-in. Hence

$M_{cx} = M_y = 47.55$ kip-in.

$F_c = \dfrac{M_{cx}}{S_f} = 45$ ksi

From section B, H_b removed,

$$I_{xp} = 1.321 \text{ in.}^4 \qquad\qquad \text{(Eq. C3.1.2-2)}$$

$S_{x,\text{net min}} = \dfrac{I_{xp}}{a'/2} = $ elastic section modulus of net section

$$= \dfrac{1.321}{3/2} = 0.881 \text{ in.}^3$$

From RMI Specification—Section 4.2.2

$$S_{cx} = \left(1 - \left(\frac{1-Q}{2}\right)\left(\frac{F_c}{F_y}\right)^Q\right)S_{x,\text{net min}}$$

$$= \left(1 - \left(\frac{1-0.82}{2}\right)(1)^{0.82}\right)0.881 \text{ in.}^3$$

$$= 0.91 \times 0.881 \text{ in.}^3 = 0.801 \text{ in.}^3$$

$$M_{nx} = S_{cx}F_c = 0.881 \times 45 \text{ kip-in.} = 36.06 \text{ kip-in.}$$

(Eq. C3.1.2-1)

G. Combined Bending and Compression (LRFD Method)

AISI Specification—Section C5.2.2

$$\frac{P_u}{\phi_c P_n} + \frac{C_{mx}M_{ux}}{\phi_b M_{nx}\alpha_x} \leqslant 1.0$$

From section E,

$$P_n = 19.15 \text{ kips}$$
$$\phi_c = 0.85$$

From section D,

$$\sigma_{ex} = 37.89 \text{ ksi}$$
$$P_{Ex} = A_g\sigma_{ex} = 1.046 \times 37.89 = 39.63 \text{ kips}$$
$$\alpha_x = 1 - \frac{P_u}{P_{Ex}} \qquad \text{(Eq. C5.2.2-4)}$$

From section F,

$$M_{nx} = 36.06 \text{ kip-in.}$$
$$\phi_b = 0.9$$

For compression members in frames subject to joint translation (side sway)

$$\bar{C}_{mx} = 0.85$$

Hence, C5.2.2-1 of the AISI Specification becomes

$$\frac{P_u}{0.85 \times 19.15} + \frac{0.85 M_{ux}}{0.9 \times 36.06 \times (1 - P_u/39.63)} \leqslant 1$$

where P_u is in kips and M_{ux} is in kip-in.

H. Horizontal Loads and Frame Analysis (LRFD Method)

RMI Specification—Section 2.5.1

$$H = 0.015 P_u$$

where P_u is the factored dead load and factored product load

$$M_{ux} = 0.015 P_u \times 60/2 \qquad \text{(assuming points of contra-flexure at center of beams and columns)}$$

$$= 0.45 P_u \text{ kip-in.}$$

Assume $P_u = 12.7$ kips:

$$\frac{12.7}{0.85 \times 19.15} + \frac{0.85 \times 0.45 \times 12.7}{0.90 \times 36.06 \times (1 - 12.7/39.63)}$$

$$= 0.780 + 0.220 = 1.00$$

Hence, factored design load in upright is 12.7 kips.

REFERENCES

11.1 Rack Manufacturers Institute, Specification for the Design, Testing and Utilization of Steel Storage Racks, Rack Manufacturers Institute, Charlotte, NC, 1997.

11.2 Rack Manufacturers Institute, Specification for the Design, Testing and Utilization of Steel Storage Racks, Materials Handling Institute, Chicago, IL, 1990.

11.3 Centre for Advanced Structural Engineering, Program THIN-WALL, Users Manual, Version 1.2, School of Civil and Mining Engineering, University of Sydney, 1996.

12

Direct Strength Method

12.1 INTRODUCTION

The design methods used throughout this book to account
for local and distortional buckling of thin-walled members
in compression and bending are based on the effective
width concept for stiffened and unstiffened elements introduced in Chapter 4. The effective width method is an
elemental method since it looks at the elements forming a
cross section in isolation. It was originally proposed by Von
Karman (Ref. 4.4) and calibrated for cold-formed members
by Winter (Refs. 4.5 and 4.6). It was initially intended to
account for local buckling but has been extended to distortional buckling of stiffened elements with an intermediate
stiffener in Section B4.1 and edge-stiffened elements in
Section B4.2 of the AISI Specification. It accounts for
postbuckling by using a reduced (effective) plate width at
the design stress.

As sections become more complex with additional edge and intermediate stiffeners, the computation of the effective widths becomes more complex. Interaction between the elements also occurs so that consideration of the elements in isolation is less accurate. To overcome these problems, a new method has been developed by Schafer and Peköz (Ref. 12.1), called the *direct strength method*. It uses elastic buckling solutions for the entire member rather than the individual elements, and strength curves for the entire member.

The method had its genesis in the design method for distortional buckling of thin-walled sections developed by Hancock, Kwon, and Bernard (Ref. 12.2). This method was incorporated in the Australian/New Zealand Standard for Cold-Formed Steel Structures (Ref. 1.4) and has been used successfully to predict the distortional buckling strength of flexural and compression members since 1996. However, the direct strength method goes one step further and assumes that local buckling behavior can also be predicted by using the elastic local buckling stress of the whole section with an appropriate strength design curve for local instability. The method has the advantage that calculations for complex sections are very simple, as demonstrated in the examples following, provided elastic buckling solutions are available.

12.2 ELASTIC BUCKLING SOLUTIONS

The finite strip method of buckling analysis described in Chapter 3 provides elastic buckling solutions suitable for use with the direct strength method and serves as a useful starting point. For the lipped channels shown in compression in Figure 3.6 and in bending in Figure 3.12, there are three basic buckling modes:

1. Local
2. Flange-distortional
3. Overall

The *local mode* involves only plate flexure in the buckling mode with the line junctions between adjacent plates remaining straight. It can occur for lipped channels, as shown in Figures 3.6 and 3.12, or unlipped channels, as shown at point A in Figure 3.3. The mode has a strong postbuckling reserve and occurs at short half-wavelengths.

The *flange-distortional mode* involves membrane bending of the stiffening elements such as the edge stiffeners shown in Figures 3.6 and 3.12. Plate flexure also occurs so that the mode has a moderate postbucking reserve. It occurs at intermediate half-wavelengths.

The *overall mode* involves translation of cross sections of the member without section distortion. It may consist of simple column (Euler) buckling as shown at point C in Figure 3.3, torsional-flexural buckling as shown at point D in Figures 3.3 and 3.6 for columns, or lateral buckling as shown at point C in Figure 3.12 for beams. It occurs at longer half-wavelengths and has very little postbuckling reserve. The overall mode may be restrained by bracing or sheathing as shown at point D in Figure 3.12. The resulting lateral-distortional mode at longer half-wavelengths is not regarded as a distortional mode in the direct strength method but should be treated as a type of hybrid overall mode.

The direct strength method uses the following solutions. For *local buckling*, the buckling stress F_{crl} is the minimum point for the local mode on a graph of stress versus half-wavelength as shown in Figures 3.3, 3.6, and 3.12. The buckling stress may be replaced by a load for compression or by a moment for bending to simplify the calculations. The interaction between the different elements is accounted for so that simple elastic local buckling coefficients, such as $k = 4$ (see Table 4.1), for a simple stiffened element in compression no longer apply. Elastic buckling solutions for simple sections of the type given by Bulson (Ref. 4.2) could be used instead of the finite strip method.

For *flange-distortional buckling*, the buckling stress F_{crd} is the minimum point for the flange-distortional mode

377

on a graph of buckling stress versus half-wavelength as shown at point B in Figures 3.6 or 3.12. The buckling stress may be replaced by a load for compression or a moment for bending to simplify the calculations. The interaction between the different elements is automatically accounted for as it should be for such complex modes. Elastic buckling solutions for edge-stiffened sections are given for compression members in Lau and Hancock (Ref. 3.8) and for flexural members in Schafer and Peköz (Ref. 12.3) and Hancock (Ref. 12.4), and can be used instead of the finite strip method.

For the *overall modes*, the elastic buckling stresses (F_e) predicted by the simple formulae in Chapter C of the AISI Specification are used. The reason for using the AISI Specification rather than the finite strip analysis is that boundary conditions other than simple supports are not accounted for in the finite strip method. Further, for flexural members, moment gradient cannot be accounted for in the finite strip method. By comparison, the design formulae in the AISI Specification can easily take account of end boundary conditions using effective length factors and moment gradient using C_b factors as described in Section C3.1.2 of the AISI Specification.

12.3 STRENGTH DESIGN CURVES

12.3.1 Local Buckling

Local buckling direct strength curves for individual elements have already been discussed and were included in Figure 4.5 for stiffened compression elements and in Figure 4.6 for unstiffened compression elements. The limiting stress on the full plate element has been called the effective design stress in these figures. The concept is that at plate failure, either the effective width can be taken to be at yield or the full width can be taken to be at the effective design stress. This concept can be generalized for sections

378

so that a limiting stress on the gross section, either in compression or bending, can be defined for the local buckling limit state. The resulting method is the direct strength method. The research of Schafer and Peköz (Ref. 12.1) has indicated that the limiting stress (F_{nl}) for local buckling of a full section is given by

$$F_{nl} = F_y \qquad \text{for } \lambda_l \leq 0.776 \tag{12.1}$$

$$F_{nl} = \left(1 - 0.15\left(\frac{F_{crl}}{F_y}\right)^{0.4}\right)\left(\frac{F_{crl}}{F_y}\right)^{0.4} F_y \qquad \text{for } \lambda_l > 0.776$$

$$\tag{12.2}$$

where

$$\lambda_l = \sqrt{\frac{F_y}{F_{crl}}} \tag{12.3}$$

The 0.4 exponent in Eq. (12.2), rather than 0.5 as used in the Von Karman and Winter formulae discussed in Chapter 4, reflect a higher post-local-buckling reserve for a complete section when compared with an element. A comparison of local buckling moments in laterally braced beams is shown by the crosses (×) in Figure 12.1 compared with Eqs. (12.1)–(12.3).

The local buckling limiting stress (F_{nl}) can be multiplied with the full unreduced section area (A) for a column or full unreduced section modulus (S_f) for a beam to get the local buckling column strength (P_{nl}) or the local buckling beam strength (M_{nl}), respectively. The resulting alternative formulations of Eqs. (12.1)–(12.3) in terms of load or moment for compression or bending respectively are

For compression

$$P_{nl} = P_y \qquad \text{for } \lambda_l \leq 0.776 \tag{12.4}$$

$$P_{nl} = \left(1 - 0.15\left(\frac{P_{crl}}{P_y}\right)^{0.4}\right)\left(\frac{P_{crl}}{P_y}\right)^{0.4} P_y \quad \text{for } \lambda_l > 0.776 \tag{12.5}$$

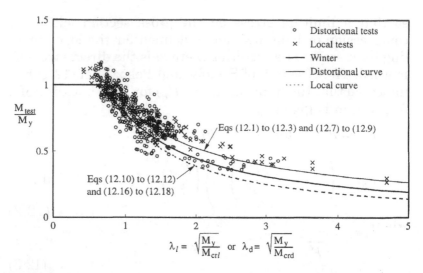

FIGURE 12.1   Laterally braced beams: bending data.

where

$$\lambda_l = \sqrt{\frac{P_y}{P_{crl}}} \tag{12.6}$$

For bending

$$M_{nl} = M_y \qquad \text{for } \lambda_l \leq 0.776 \tag{12.7}$$

$$M_{nl} = \left(1 - 0.15\left(\frac{M_{crl}}{M_y}\right)^{0.4}\right)\left(\frac{M_{crl}}{M_y}\right)^{0.4} M_y \text{ for } \lambda_l > 0.776 \tag{12.8}$$

where

$$\lambda_l = \sqrt{\frac{M_y}{M_{crl}}} \tag{12.9}$$

In these equations, the loads have been derived from the stresses by multiplying by the full unreduced section area (A) throughout, and the moments have been derived from the stresses by multiplying by the full unreduced section modulus (S_f) throughout.

12.3.2 Flange-Distortional Buckling

Flange-distortional buckling direct strength curves were developed for sections in both compression and bending by Hancock, Kwon, and Bernard (Ref. 12.2) and are shown in Figure 12.2. Research has indicated that the limiting stress (F_{nd}) for distortional buckling of a full section is given by Eqs. (12.10)–(12.12). A comparison of distortional buckling loads and moments is shown in Figure 12.2 compared with Eqs. (12.10) to (12.12).

$$F_{nd} = F_y \qquad \text{for } \lambda_d \leq 0.561 \tag{12.10}$$

$$F_{nd} = \left(1 - 0.25\left(\frac{F_{crd}}{F_y}\right)^{0.6}\right)\left(\frac{F_{crd}}{F_y}\right)^{0.6} F_y \quad \text{for } \lambda_d > 0.561 \tag{12.11}$$

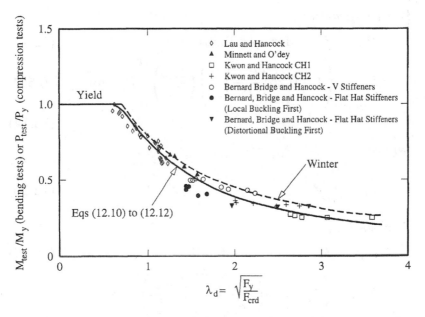

FIGURE 12.2 Comparison of distortional buckling test results with design curves.

where

$$\lambda_d = \sqrt{\frac{F_y}{F_{crd}}} \tag{12.12}$$

The 0.6 exponent in Eq. (12.11), rather than 0.4 as used for local buckling strength in Eqs. (12.1) to (12.3), reflects a lower postbuckling reserve for a complete section in the distortional mode than the local mode.

The distortional buckling limiting stress (F_{nd}) can be multiplied with the full unreduced section area (A) for a column or full unreduced section modulus (S_f) for a beam to get the distortional buckling column strength (P_{nd}) or the distortional buckling beam strength (M_{nd}). The resulting alternative formulations in Eqs. (12.10)–(12.12) in terms of load or moment for compression or bending, respectively, are

For compression

$$P_{nd} = P_y \quad \text{for } \lambda_d \leq 0.561 \tag{12.13}$$

$$P_{nd} = \left(1 - 0.25\left(\frac{P_{crd}}{P_y}\right)^{0.6}\right)\left(\frac{P_{crd}}{P_y}\right)^{0.6} P_y \quad \text{for } \lambda_d > 0.561 \tag{12.14}$$

where

$$\lambda_d = \sqrt{\frac{P_y}{P_{crd}}} \tag{12.15}$$

For bending

$$M_{nd} = M_y \quad \text{for } \lambda_d \leq 0.561 \tag{12.16}$$

$$M_{nd} = \left(1 - 0.25\left(\frac{M_{crd}}{M_y}\right)^{0.6}\right)\left(\frac{M_{crd}}{M_y}\right)^{0.6} M_y$$

$$\text{for } \lambda_d > 0.561 \tag{12.17}$$

where

$$\lambda_d = \sqrt{\frac{M_y}{M_{crd}}} \qquad (12.18)$$

In these equations, the loads have been derived from the stresses by multiplying by the full unreduced section area (A) throughout, and the moments have been derived from the stresses by multiplying by the full unreduced section modulus (S_f) throughout.

12.3.3 Overall Buckling

The overall buckling strength design curves are those specified in Section C of the AISI Specification. For compression members, they are described in Section 7.4 of this book and are shown in Figure 7.3. The design stress (F_n) for compression members is as specified in Section C4 of the AISI Specification. For flexural members, they are described in Section 5.2.3 of this book and shown in Figure 5.8. The design stress (F_c) for flexural members is specified in Section C3.1.2 of the AISI Specification.

The design stress (F_n) for compression members can be multiplied with the full unreduced section area (A) to give the inelastic long-column buckling load (P_{ne}):

$$P_{ne} = AF_n \qquad (12.19)$$

Similarly, the design stress (F_c) for flexural members can be multiplied with the full unreduced section modulus (S_f) to give the inelastic lateral buckling moment (M_{ne}):

$$M_{ne} = S_f F_c \qquad (12.20)$$

12.4 DIRECT STRENGTH EQUATIONS

The direct strength method allows for the interaction of local and overall buckling of columns by a variant of the unified approach described in Chapter 7 for compression members. This is achieved simply by replacing P_y in Eqs.

(12.4)–(12.6) for local buckling by P_{ne} from Eq. (12.19). The resulting limiting load (P_{nl}) as given by Eqs. (12.21)–(12.23) accounts for the interaction of local buckling with overall column buckling since the limiting load is P_{ne} rather than P_y. A comparison of the test strengths for failure in the local mode is shown by the crosses (×) in Figure 12.3 compared with Eqs. (12.21)–(12.23).

$$P_{nl} = P_{ne} \qquad \text{for } \lambda_l \le 0.776 \qquad\qquad (12.21)$$

$$P_{nl} = \left(1 - 0.15\left(\frac{P_{crl}}{P_{ne}}\right)^{0.4}\right)\left(\frac{P_{crl}}{P_{ne}}\right)^{0.4} P_{ne} \qquad \text{for } \lambda_l > 0.776$$

$$(12.22)$$

where

$$\lambda_l = \sqrt{\frac{P_{ne}}{P_{crl}}} \qquad\qquad (12.23)$$

A similar set of equations can be derived for the interaction of local and lateral buckling of beams, using a

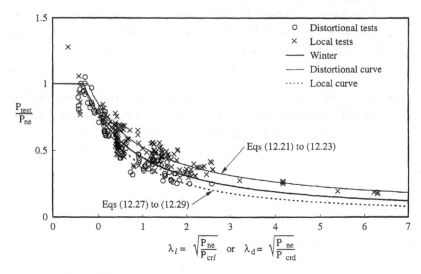

FIGURE 12.3   Pin-ended columns: compression data.

variant of the unified approach described in Chapter 5 for flexural members. This is achieved simply by replacing M_y in Eqs. (12.7) –(12.9) for local buckling by M_{ne} from Eq. (12.20). The resulting limiting moment (M_{nl}) as given by Eqs. (12.24)–(12.26) accounts for the interaction of local buckling with lateral buckling since the limiting moment is M_{ne} rather than M_y.

$$M_{nl} = M_{ne} \quad \text{for } \lambda_l \leq 0.776 \quad (12.24)$$

$$M_{nl} = \left(1 - 0.15\left(\frac{M_{crl}}{M_{ne}}\right)^{0.4}\right)\left(\frac{M_{crl}}{M_{ne}}\right)^{0.4} M_{ne}$$

$$\text{for } \lambda_l > 0.776 \quad (12.25)$$

where

$$\lambda_l = \sqrt{\frac{M_{ne}}{M_{crl}}} \quad (12.26)$$

The direct strength method also allows for the interaction of distortional and overall buckling by a further variant of the unified approach. This is achieved simply by replacing P_y in Eqs. (12.13)–(12.15) for distortional buckling by P_{ne} from Eq. (12.19). The resulting limiting load (P_{nd}) as given by Eqs. (12.27)–(12.29) accounts for the interaction of distortional buckling with overall column buckling since the limiting load is P_{ne} rather than P_y. A comparison of the test strengths for failure in the distortional mode is shown by the circles (O) in Figure 12.3 compared with Eqs. (12.27)–(12.29).

$$P_{nd} = P_{ne} \quad \text{for } \lambda_d \leq 0.561 \quad (12.27)$$

$$P_{nd} = \left(1 - 0.25\left(\frac{P_{crd}}{P_{ne}}\right)^{0.6}\right)\left(\frac{P_{crd}}{P_{ne}}\right)^{0.6} P_{ne}$$

$$\text{for } \lambda_d > 0.561 \quad (12.28)$$

where

$$\lambda_d = \sqrt{\frac{P_{ne}}{P_{crd}}} \qquad (12.29)$$

A similar set of equations can be derived for the interaction of distortional and lateral buckling of beams, using a further variant of the unified approach. This is achieved simply by replacing M_y in Eqs. (12.16)–(12.18) for distortional buckling by M_{ne} from Eq. (12.20). The resulting limiting moment (M_{nd}) as given by Eqs. (12.30)–(12.32) accounts for the interaction of local buckling with lateral buckling since the limiting moment is M_{ne} rather than M_y.

$$M_{nd} = M_{ne} \qquad \text{for } \lambda_d \le 0.561 \qquad (12.30)$$

$$M_{nd} = \left(1 - 0.25\left(\frac{M_{crd}}{M_{ne}}\right)^{0.6}\right)\left(\frac{M_{crd}}{M_{ne}}\right)^{0.6} M_{ne}$$

$$\text{for } \lambda_d > 0.561 \quad (12.31)$$

where

$$\lambda_d = \sqrt{\frac{M_{ne}}{M_{crd}}} \qquad (12.32)$$

The nominal member strength is the lesser of P_{nl} and P_{nd} for compression members and of M_{nl} and M_{nd} for flexural members. It has been suggested that the same resistance factors and factors of safety apply as for normal compression and flexural member design.

12.5 EXAMPLES

12.5.1 Lipped Channel Column (Direct Strength Method)

Problem

Determine the nominal axial strength (P_n) of the lipped channel in Example 7.6.3 by using the direct strength method. The geometry is shown in Figure 7.9 and the dimensions in Example 7.6.3.

Solution

A. **Compute the Elastic Local and Distortional Buckling Stresses Using the Finite Strip Method**

Program THIN-WALL (Ref. 11.3)

The elastic local (F_{crl}) and distortional buckling stresses are shown in Figure 12.4.

$$F_{crl} = 31.8 \text{ ksi at 3.5-in. half-wavelength}$$

$$F_{crd} = 34.4 \text{ ksi at 24-in. half-wavelength}$$

$$A = 0.645 \text{ in}^2.$$

$$P_{crl} = AF_{crl} = 0.645 \times 31.8 = 20.51 \text{ kips}$$

$$P_{crd} = AF_{crd} = 0.645 \times 34.4 = 22.2 \text{ kips}$$

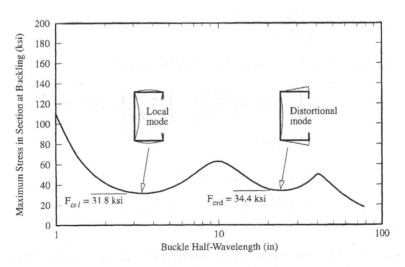

FIGURE 12.4 Lipped channel in compression.

Chapter 12

B. Compute the Inelastic Long Column Buckling Load (P_{ne})

$$F_e = 54.13 \text{ ksi} \qquad\qquad \text{(from Ex. 7.6.3B)}$$
$$F_n = 31.78 \text{ ksi} \qquad\qquad \text{(from Ex. 7.6.3B)}$$

The buckling mode is torsional-flexural:

$$P_{ne} = AF_n = 0.645 \times 31.78 = 20.50 \text{ kips}$$

C. Compute the Local Buckling Strength (P_{nl})

$$\lambda_l = \sqrt{\frac{P_{ne}}{P_{crl}}} = \sqrt{\frac{20.50}{20.51}} = 1.0 > 0.776$$

Since $\lambda_l > 0.776$,   use Eq. (12.22) to get

$$P_{nl} = \left(1 - 0.15\left(\frac{20.51}{20.50}\right)^{0.4}\right)\left(\frac{20.51}{20.50}\right)^{0.4} 20.50$$

$$= 17.43 \text{ kips}$$

D. Compute the Distortional Buckling Strength (P_{nd})

$$\lambda_d = \sqrt{\frac{P_{ne}}{P_{crd}}} = \sqrt{\frac{20.5}{22.2}} = 0.961 > 0.561$$

Since $\lambda_d > 0.561$,   use Eq. (12.28) to get

$$P_{nd} = \left(1 - 0.25\left(\frac{22.2}{20.5}\right)^{0.6}\right)\left(\frac{22.2}{20.5}\right)^{0.6} 20.50$$

$$= 15.79 \text{ kips}$$

E. Nominal Member Strength (P_n)

P_n is the lesser of P_{nl} and P_{nd}, so
$$P_n = 15.79 \text{ kips}$$

This can be compared with 15.75 kips in Example 7.6.3 obtained by using the effective width method.

12.5.2 Simply Supported C-Section Beam

Problem

Determine the design load on the C-section beam in Example 5.6 by using the direct strength method. The section geometry is shown in Figure 4.12 and the beam geometry in Figure 5.20. The section dimensions are given in Example 4.6.3 and the beam dimensions in Figure 5.20.

Solution

A. Compute the Elastic Local and Distortional Buckling Stresses using the Finite Strip Method

Program THIN-WALL (Ref. 11.3)

The elastic local (F_{crl}) and distortional buckling stresses are shown in Figure 12.5.

$$F_{crl} = 45.8 \text{ ksi at 4.5-in. half-wavelength}$$

$$F_{crd} = 40.7 \text{ ksi at 25-in. half-wavelength}$$

$$S_f = 2.049 \text{ in}^3. \qquad \text{(from Ex. 5.6.1)}$$

$$M_{crl} = S_f F_{crl} = 2.049 \times 45.8 = 93.8 \text{ kip-in.}$$

$$M_{crd} = S_f F_{crd} = 2.049 \times 40.7 = 83.4 \text{ kip-in.}$$

B. Compute the Inelastic Lateral Buckling Moment (M_{ne})

$$F_c = 24.5 \text{ ksi} \qquad \text{(see Ex. 5.6.1)}$$

$$M_{ne} = S_f F_c = 2.049 \times 24.5 = 50.2 \text{ kip-in.}$$

C. Compute the Local Buckling Strength (M_{nl})

$$\lambda_l = \sqrt{\frac{M_{ne}}{M_{crl}}} = \sqrt{\frac{50.2}{93.8}} = 0.732 < 0.776$$

Since $\lambda_l < 0.776$, use Eq. (12.24) to get

$$M_{nl} = M_{ne} = 50.2 \text{ kip-in.}$$

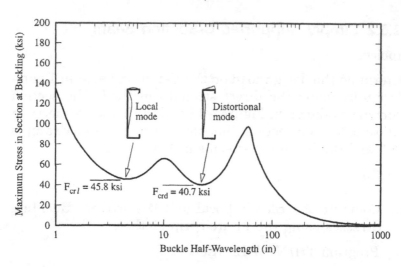

FIGURE 12.5   C-section in bending.

D.   Compute the Distortional Buckling Strength (M_{nd})

$$\lambda_d = \sqrt{\frac{M_{ne}}{M_{crd}}} = \sqrt{\frac{50.2}{83.4}} = 0.776 > 0.561$$

Since $\lambda_d > 0.561$, use Eq. (12.31) to get

$$M_{nd} = \left(1 - 0.25\left(\frac{83.4}{50.2}\right)^{0.6}\right)\left(\frac{83.4}{50.2}\right)^{0.6} 50.2$$

$$= 45.0 \text{ kip-in.}$$

E.   Nominal Member Strength (M_n)

M_n is the lesser of M_{nl} and M_{nd}, so $M_n = 45.0$ kip-in.

This can be compared with 47.7 kip-in. in Example 5.6.1 using the effective width method.

REFERENCES

12.1 Schafer, B.W. and Peköz, T., Direct Strength Prediction of Cold-Formed Steel Members Using Numerical Elastic Buckling Solutions, Thin-Walled Structures, Research and Development, Eds. Shanmugan, N.E., Liew, J.Y.R., and Thevendran, V., Elsevier, 1998, pp. 137–144 (also in Fourteenth International Specialty Conference on Cold-Formed Steel Structures, St Louis, MO, Oct. 1998).

12.2 Hancock, G.J., Kwon, Y.B., and Bernard, E.S., Strength design curves for thin-walled sections undergoing distortional buckling, J. Constr. Steel Res., Vol. 31, Nos. 2/3, 1994, pp. 169–186.

12.3 Schafer, B.W. and Peköz, T., Laterally braced cold-formed steel members with edge stiffened flanges, Journal of Structural Engineering, ASCE, 1999, Vol. 125, No. 2, pp. 118–127.

12.4 Hancock, G.J., Design for distortional buckling of flexural members, Thin-Walled Structures, Vol. 20, 1997, pp. 3–12.

Index